AF213696

Applied Condition Monitoring

Volume 14

Series editors

Mohamed Haddar, National School of Engineers of Sfax, Tunisia

Walter Bartelmus, Wrocław University of Technology, Poland

Fakher Chaari, National School of Engineers of Sfax, Tunisia
e-mail: fakher.chaari@gmail.com

Radoslaw Zimroz, Wrocław University of Technology, Poland

The book series Applied Condition Monitoring publishes the latest research and developments in the field of condition monitoring, with a special focus on industrial applications. It covers both theoretical and experimental approaches, as well as a range of monitoring conditioning techniques and new trends and challenges in the field. Topics of interest include, but are not limited to: vibration measurement and analysis; infrared thermography; oil analysis and tribology; acoustic emissions and ultrasonics; and motor current analysis. Books published in the series deal with root cause analysis, failure and degradation scenarios, proactive and predictive techniques, and many other aspects related to condition monitoring. Applications concern different industrial sectors: automotive engineering, power engineering, civil engineering, geoengineering, bioengineering, etc. The series publishes monographs, edited books, and selected conference proceedings, as well as textbooks for advanced students.

More information about this series at http://www.springer.com/series/13418

Tomasz Barszcz

Vibration-Based Condition Monitoring of Wind Turbines

Springer

Tomasz Barszcz
AGH University of Science and Technology
Kraków, Poland

ISSN 2363-698X ISSN 2363-6998 (electronic)
Applied Condition Monitoring
ISBN 978-3-030-05969-9 ISBN 978-3-030-05971-2 (eBook)
https://doi.org/10.1007/978-3-030-05971-2

Library of Congress Control Number: 2018964244

© Springer Nature Switzerland AG 2019
This work is subject to copyright. All rights are reserved by the Publisher, whether the whole or part of the material is concerned, specifically the rights of translation, reprinting, reuse of illustrations, recitation, broadcasting, reproduction on microfilms or in any other physical way, and transmission or information storage and retrieval, electronic adaptation, computer software, or by similar or dissimilar methodology now known or hereafter developed.
The use of general descriptive names, registered names, trademarks, service marks, etc. in this publication does not imply, even in the absence of a specific statement, that such names are exempt from the relevant protective laws and regulations and therefore free for general use.
The publisher, the authors, and the editors are safe to assume that the advice and information in this book are believed to be true and accurate at the date of publication. Neither the publisher nor the authors or the editors give a warranty, express or implied, with respect to the material contained herein or for any errors or omissions that may have been made. The publisher remains neutral with regard to jurisdictional claims in published maps and institutional affiliations.

This Springer imprint is published by the registered company Springer Nature Switzerland AG
The registered company address is: Gewerbestrasse 11, 6330 Cham, Switzerland

To my wife, Katarzyna

To my wife Kamonyak

Preface

Wind turbines have become very popular in the landscape in recent years. Their appearance triggered a number of discussions concerning their impact on the environment, their efficiency, potential threat to animals, and plethora of other subjects. For a mechanical engineer, however, wind turbines are complex machines operating in very challenging conditions. Cost structure of operating a wind farm is very different from a thermal plant. In a wind farm, a majority of costs must be paid before turbines start to generate income. In other words, capital expenditures (CAPEX) are very high and operating expenditures (OPEX) can be low, as the energy source is free of charge. In such a situation, an uninterrupted readiness to generate electricity is a key requirement. An engineer will say that we require high availability. For machines with gearboxes (which make up a majority of all wind turbines), the most vulnerable part of a wind turbine is its drivetrain, i.e., a main bearing, a gearbox, and a generator. A single unplanned exchange of a gearbox can cost several hundred thousand euros, and methods which can tell us whether the gearbox is in a good shape are very important to the users.

This book explains how the technical state of a wind turbine drivetrain can be assessed, based on the vibration analysis of its mechanical vibration. After the introduction, Chap. 1 starts with the description of vibration signals used for monitoring drivetrains and presents its key features. The signal processing methods, including the advanced ones, like signal resampling and signal envelope, are described in Chap. 2. This chapter puts great importance to the fact that turbines generate electricity in response to the wind. It is a fundamental cause why they work in highly varying operational conditions.

Vibration-based condition monitoring has become an important branch of the market. There are several monitoring devices available to potential users. They vary greatly in features and applications. Chapter 3 describes types of devices used for condition monitoring purposes, ranging from vibration sensors, through supervisory control and data acquisition (SCADA), to portable data analyzers and online condition monitoring systems.

Vibration analysis is a vibrant research field, in which new methods are introduced to help to process vibration data more accurately. Chapters 4 and 5 constitute the main part of this book and are dedicated to such new methods. It is accompanied by real case studies, in which advanced signal processing methods were used to detect failures of gearboxes and bearings of wind turbines.

This book is intended for researchers in the field of vibration signal analysis interested in wind turbines. It will provide them with an in-depth understanding of the most recent research achievements in this domain. It can also be useful for practitioners active in the field of wind turbine condition monitoring, help them in extending their knowledge in the field, and give examples of equipment available on the market. Finally, it will be interesting to graduate students who would like to extend their knowledge into the field of vibration analysis. The information contained in this book will also be valuable to those interested in condition monitoring of other machines working in varying operational conditions, like airplanes, helicopters, vehicles, mining equipment.

Kraków, Poland Tomasz Barszcz

Acknowledgements

During writing this book, I had a privilege of working with wonderful people, who contributed to the formation of this work. First of all, I'd like to thank Adam Jablonski and Jacek Urbanek, who completed their Ph.D. under my supervision. They both combined curiosity and imagination with hard work. The journey with them into the land of vibration signal processing was a very productive and fortunate period of my life. I also would like to thank my mentors and colleagues who accompanied me on my scientific trip. I have special thanks to Prof. J. Antoni, Prof. W. Bartelmus, Dr. N. Martin, Prof. B. Randall, Prof. T. Uhl, Prof. A. Wylomanska, and Prof. R. Zimroz. We had so many stimulating discussions, often resulting in innovative ideas and common work. My research in the field of wind power generation was in large part initiated by real problems of real machines. I would like to say thanks to my colleagues and friends from companies I was privileged to work with: PGE, ENEA, Energa, RP Global, GE, Alstom, Siemens, SEACOM, EC Systems, and AMC VIBRO.

Writing of this book took a long time and was often delayed to many other duties. I would like to give a very special thanks to my wife, Katarzyna. She motivated and encouraged me in this effort. Moreover, many daily chores were taken out of my head so I had enough time to finish this volume.

Contents

Abbreviations

ADC	Analog Digital Converter
AE	Acoustic Emission
AI	Artificial Intelligence
AM	Amplitude Modulation
AZT	Allianz Zentrum fuer Technik
BPFI	Ball Pass Frequency of the Inner Ring
BPFO	Ball Pass Frequency of the Outer Ring
BSF	Ball Spin Frequency
BW	Bandwidth
CAPEX	Capital Expenditures
CF	Crest Factor
CF	Center Frequency
CM	Condition Monitoring
CMMNO	Condition Monitoring of Machinery in Non-stationary Operations
CMS	Condition Monitoring System
CMS	Cyclic Modulation Spectrum
CPU	Central Processing Unit
DFT	Discrete Fourier Transform
EK	Excess Kurtosis
ESK	Envelope Spectrum Kurtosis
FEM	Finite Element Model
FFT	Fast Fourier Transform
FIR	Finite Impulse Response
FM	Frequency Modulation
FRF	Frequency Response Function
FTF	Fundamental Train Frequency
GL	Germanisher Lloyd
GMF	Gear Mesh Frequency

HFRT	High Frequency Resonance Technique
HP	High Pass
HT	Hunting Tooth
IAS	Instantaneous Angular Speed
ICPCP	Instantaneous Circular Pitch Cyclic Power
IEPE	Integrated Electronics Piezo-Electric
IMD	Integrated MID
IPS	Instantaneous Power Spectrum
ISO	International Standardization Organization
LP	Low Pass
LSCh	Load Susceptibility
MCSA	Motor Current Signature Analysis
MEMS	Micro Electro Mechanical Systems
MID	Modulation Intensity Distribution
MIF	Modulation Intensity Factor
MIMO	Multi Input Multi Output
NEA	Narrowband Envelope Analysis
NREL	National Renewable Energy Laboratory
OFB	Optimum Frequency Band
OPEX	Operational Expenditures
PC	Personal Computer
PDF	Probability Distribution Function
PP	Peak–Peak
PSC	Product of two Spectral Correlations
PSCoh	Product of two Spectral Coherences
PSD	Power Spectrum Density
REB	Rolling Element Bearing
RMS, rms	Root Mean Square
RPM, rpm	Rotations Per Minute
RUL	Remaining Useful Life
SCADA	Supervisory Control and Data Acquisition
SCD	Spectral Correlation Density
SCoh	Spectral Coherence
SCohD	Spectral Coherence Density
SK	Spectral Kurtosis
SNR	Signal-to-Noise Ratio
SOI	Signal of Interest
SP	Spectrogram
STFT	Short Time Fourier Transform
SVM	Support Vector Machine
TC	Technical Committee
TSA	Time Synchronous Averaging

VDI	Verein Deutscher Ingenieure
VOC	Varying Operational Conditions
WT	Wind Turbine
WTAS	Wind Turbine Analysis System
ZP	Zero–Peak

Symbols

N, L	Integer number
t	Continuous time
k	Discrete time
τ	Time lag
f	Frequency
α	Cyclic frequency
T	Period
TH	Threshold value
$x, y, \ldots$	Time signals
$X, Y, \ldots$	Fourier transforms of signals
df	Frequency resolution
ω	Cyclic frequency
φ	Angle
h	Impulse response function
H	Frequency response function
$1X, 2X, \ldots$	1, 2, … harmonics
M	Amplitude
c	Complex coefficient
$hann$	Profile of the window function
$E(), \langle\cdot\rangle$	Averaging operators
$p()$	Probability density
μ	Mean value
σ	Standard deviation
sup	Supremum
K	Kurtosis
R	Autocorrelation
$\mathcal{P}_0$	Averaging operator
$x_{\Delta f}(t; f)$	$x(t)$ filtered through the frequency band with center frequency f of width Δf
$\mathcal{P}_\alpha\{\cdot\}$	Operator extracting the periodic component at a frequency α
$\mathcal{P}\{\cdot\}$	Operator extracting all cyclic frequencies

$P_x(t, f;\ \Delta f)$	Instantaneous power spectrum
$P_x^{\propto}(f;\ \Delta f)$	Cyclic power spectrum
$SC_x^{\propto}(f)$	Spectral correlation density
$\gamma_x^{\alpha}(f)$	Spectral coherence density
$MID(\cdot)_{\Delta f}(f, \alpha)$	Modulation intensity distribution with modulation intensity factor (.)
$IMD(\cdot)_{f_1}^{f_2}(\alpha, \Delta f)$	Integrated MID in frequency range $f_1 \ldots f_2$

Chapter 1
Introduction

1.1 Who Should Read This Book

My career path comprises sharing my time between academia and industry. I have experienced and appreciated how much both communities are different and how much they can learn from one another. The most important field for me is vibration based condition monitoring of wind turbine drivetrains. This has become my main motivation to write a book on the topic which I hope will be useful for both researchers and practitioners.

During the past 25 years, I have come across many books on vibration signal analysis. They present the subject from many angles, but still I could not find any book devoted to wind turbine drivetrains. There are hundreds of scientific papers being published in this domain and adjacent ones. For most readers, however, it is really time consuming to browse through all the papers and pick up the ones relevant to wind turbine drivetrains. My ambition is to fill this gap by providing basic information about the wind turbines market and design, a comprehensive survey of currently used vibration analysis methods and available condition monitoring systems, finally, presenting recent research in this field.

Important aspects of condition monitoring of wind turbines are varying operational conditions of these machines. Constant wind variability causes variations in all the process parameters, primarily the rotational speed and the generated power. Since the rotational speed governs frequencies generated by all the drivetrain components, the frequency spectrum is very different from this of a constant speed machine. A varying load is another factor changing amplitudes of vibration signals. All of it needs to be included in efficient analysis of a wind turbine technical state.

This book is primarily intended for researchers in the field of vibration signal analysis interested in wind turbines and it will provide them with in-depth understanding of the most recent research achievements in this domain. I have included

© Springer Nature Switzerland AG 2019

T. Barszcz, *Vibration-Based Condition Monitoring of Wind Turbines*, Applied Condition Monitoring 14, https://doi.org/10.1007/978-3-030-05971-2_1

the most recent (as of February 2018) research results in the field. My main interest was detection of gear and rolling bearing faults with vibration signals. The book contains several case studies which provide additional understanding of the presented methods.

It can also be useful for practitioners active in the field of wind turbines condition monitoring and help them in extending their knowledge in the field. The important part of the book is the survey about the equipment available on the market. As this market is mature now, there is a wide variety of off-the-shelf products which can be applied for condition based monitoring. To achieve the best possible return on investment from the condition monitoring the key requirements toward the CM implementation process are stressed.

Last but not least, the book will also be interesting to graduate students who would like to extend their knowledge into the field of vibration analysis. The first chapters include a compendium of vibration signal analysis methods starting from basic broadband features to the advanced ones.

The information contained in this book will also be valuable to those interested in condition monitoring of other machines working in varying operational conditions, like airplanes, helicopters, vehicles, mining equipment and others. Most of the techniques presented in the book can be quickly applied in the domains mentioned above.

1.2 Types of Drivetrains

Wind turbine technology has been developed for centuries. Nevertheless, the wide application of wind power generation to a public grid has gained momentum since the 80s. During this time dozens of different machines converting wind energy into its other forms have been invented, designed and tested. The complete process involves information and experience from many fields of knowledge, e.g. aerodynamics, material science, mechanical and electrical engineering. An interested reader can find facts and data about the historical developments and physical principles in books by Hau [1] or Burton et al. [2]. In these book the most popular design is going to be considered: a horizontal axis upwind turbine. The term “upwind” depicts a turbine in which the main rotor is the part directed towards the wind.

A drivetrain itself consists of the following parts:

- main shaft,
- main bearing,
- gearbox (only in the gear design),
- coupling,
- generator.

There are also other mechanical components in the nacelle, though they are not part of a drivetrain and will not be considered here. Such mechanical subsystems are a blade pitch, a yaw and a brake. A pitch control mechanism is used in pitch

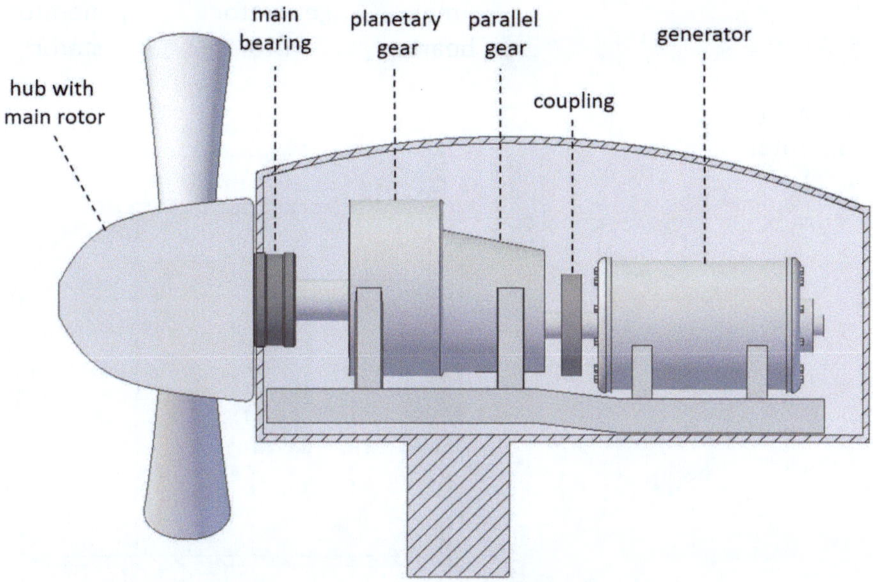

Fig. 1.1 Drivetrain of the horizontal axis wind turbine with gearbox

controlled turbines; it is mounted in a rotor hub and continuously sets the pitch angle of the main rotor blades. A yaw drive can change the azimuth and is used to direct the nacelle towards the wind. A mechanical brake is used to stop a turbine. It is typically mounted between a gearbox and a generator.

The main role of a wind turbine drivetrain is to transmit mechanical power from a main rotor to a generator. There are two fundamental setups of mechanical transmission:

- gearbox between the main rotor and the generator,
- direct drive (gearless).

The most popular design is such in which a gearbox is placed in between a rotor and a generator. The example of the most typical setup is presented in the Fig. 1.1. The main rotor with three blades is supported by the main bearing and transmits torque to the planetary gear. The planetary gear input is a plate to which the main rotor is connected. The planetary gear has three planets, with their shafts attached to the plate. The planets roll over the stationary ring and transmit torque to the sun. The sun shaft is the output of the planetary gear. Further, the sun drives the two-stage parallel gear. The parallel gear has three shafts: the slow shaft connected to the sun shaft, the intermediate shaft and the fast shaft which drives the generator. The intermediate shaft is mounted inside the parallel gear.

A gearbox is used to increase slow rotational speed of a rotor (ca. 18 rpm) to match the speed of a generator (ca. 1500 rpm for 50 Hz grid). Thus, a gearbox ratio is in the range of 80–100. As presented above, the most popular design uses a one

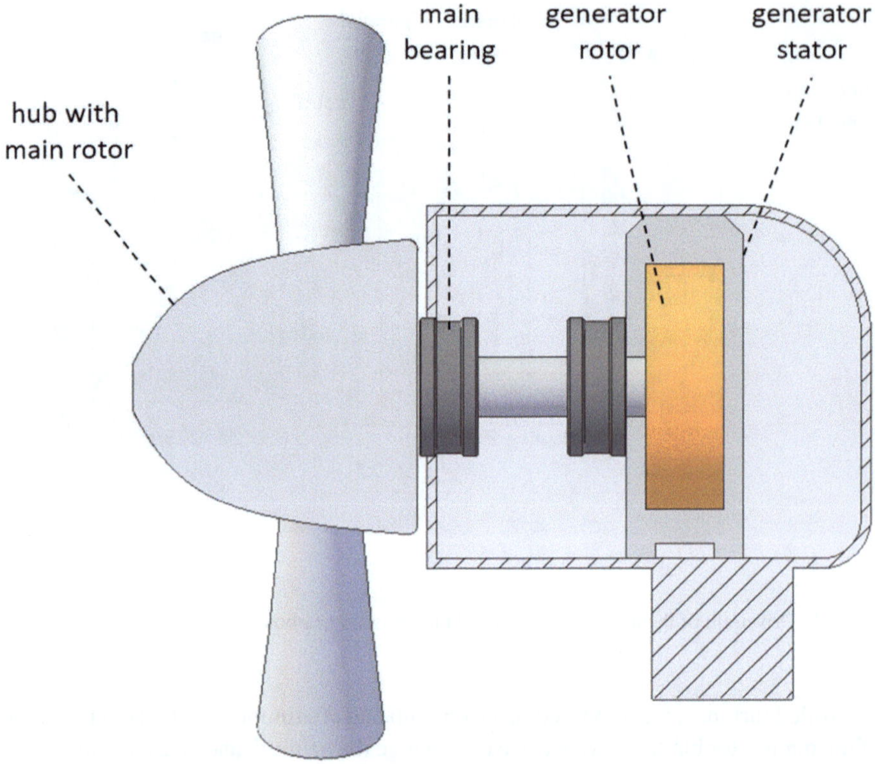

Fig. 1.2 Drivetrain of the direct drive wind turbine

stage planetary gear and a two stage parallel gear. Other gearbox designs are also manufactured, e.g. a two stage planetary gear with a one stage parallel gear. There are also gearbox designs with a three stage parallel gearing.

The main advantage of a standard design is the ability to use conventional gearboxes and generators which are used in other industries and manufactured by many independent companies. Using popular components from the market results in a relatively light and inexpensive system. Another important advantage is easy maintenance, both in terms of access to individual components as well as in availability of spare parts.

The other design, less frequent, is a direct drive (gearless) wind turbine. The main rotor is placed directly on a multi-pole generator shaft. The major problem of a gearless wind turbine is the need of a dedicated, multi-pole generator. Next, a power electronic frequency converter further increases frequency to match the grid. The example of such a design is shown in the Fig. 1.2.

As the generator rotational speed is that of the main rotor, the generator requires a high number of poles. Thus, direct drive generators have large diameters. It helps to distinguish them from the standard design. In general, direct drive machines require a

Table 1.1 Basic operational parameters of the Enercon E82 E2 wind turbine

Parameter	Value	Unit
Tower height	78–138	m
Rotor diameter	82	m
Nominal power	2000	kW
Control type	Pitch	–
Gearbox	None	–
Rotor speed	6–18	rpm
Generator speed	6–18	rpm
Weight (nacelle, rotor and blades)	Approx. 135	Mg

Table 1.2 Basic operational parameters of the Vestas V90-2 wind turbine

Parameter	Value	Unit
Tower height	80–125	m
Rotor diameter	90	m
Nominal power	2000	kW
Control type	Pitch	–
Gearbox	1 planetary + 2 parallel	–
Nominal rotor speed	18	rpm
Nominal generator speed	1500	rpm
Weight (nacelle, rotor and blades)	Approx. 108	Mg

dedicated generator resulting in heavier and more costly turbines. On the other hand, its most important advantage is the increased reliability on account of a simplified drivetrain.

Tables 1.1 and 1.2 present key parameters of two comparable 2 MW wind turbines. Please, note that the parameters given in Table 1.2 are for the 50 Hz market. For the 60 Hz market nominal generator speed for a geared type turbines is 1800 rpm.

1.3 Stall Controlled and Pitch Controlled Turbines

Another important distinction within geared wind turbines is a control type which greatly influences generated vibration signals. The main problem to solve is how to adjust turbine power to changing wind conditions in high winds. There are two main design solutions to this problem: stall control and pitch control.

Stall control takes advantage of the physical phenomenon of separation of air flow from the profile of a blade when the angle of attack is too high. In aerospace it is very dangerous as the lift force decreases suddenly and may lead to a catastrophic failure. In wind turbines, on the other hand, it is used to decrease generated power.

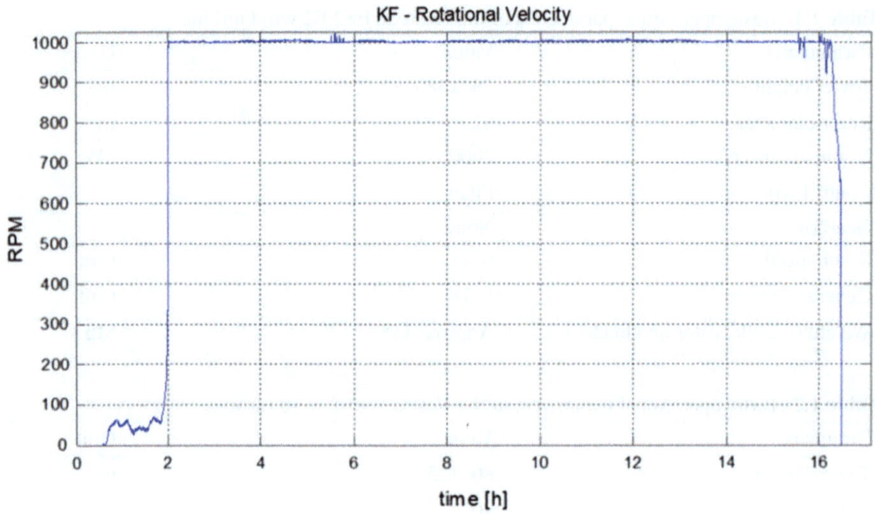

Fig. 1.3 Example of speed profile of a stall controlled wind turbine

Such a design is very simple and does not require a control system of the blades. Since it is not possible to achieve efficient generation in wide range of wind speeds, stall controlled turbines have two nominal generator speeds, most often 1000 and 1500 rpm. As long as the turbine works at a given speed, the load may vary, but the speed is almost ideally constant. It has important consequences for generated vibration signals. Every component (as it will be shown in Sect. 1.4) generates its own vibration pattern bound to so-called characteristic frequency which depends on rotational speed. If the rotational speed is constant, characteristic frequencies are constant and can be determined and analyzed by frequency spectrum. The Fig. 1.3 presents the speed profile of the stall controlled turbine. One can observe that for several hours the turbine maintained almost perfectly constant speed of 1000 rpm.

With the increasing size and power, stall turbines have lost the market to a more efficient pitch controlled design. Stall controlled ones can be found in the field, but this design rarely achieves nominal power above 1 MW. For an illustration, the Table 1.3 presents the main parameters of a stall controlled turbine.

The currently dominant model, namely the pitch control, is a design in which a pitch angle of the main rotor blades is changed depending on wind speed. The goal is to maintain the optimal operating conditions. Rotor blades operate at more efficient conditions than the stall controlled ones. Therefore, pitch control turbines are more efficient though at the cost of a complex pitch control mechanism.

As far as vibration signal analysis is concerned, continuously varying rotational speed is a major obstacle. The characteristic frequencies mentioned above do not represent constant frequency lines if the rotational speed is not constant. The Fig. 1.4 presents the speed profile of a pitch controlled turbine. There are periods when within only 3 min the rotational speed can vary from 820 to 1080 rpm. It is a change of

Table 1.3 Basic operational parameters of the NEG-Micon NM60/1000 wind turbine

Parameter	Value	Unit
Tower height	70	m
Rotor diameter	60	m
Nominal power	1070	kW
Control type	Stall	–
Gearbox	1 planetary + 2 parallel	–
Rotor speed	12 or 18	rpm
Generator speed	1000 or 1500	rpm
Weight (nacelle, rotor and blades)	Approx. 69	Mg

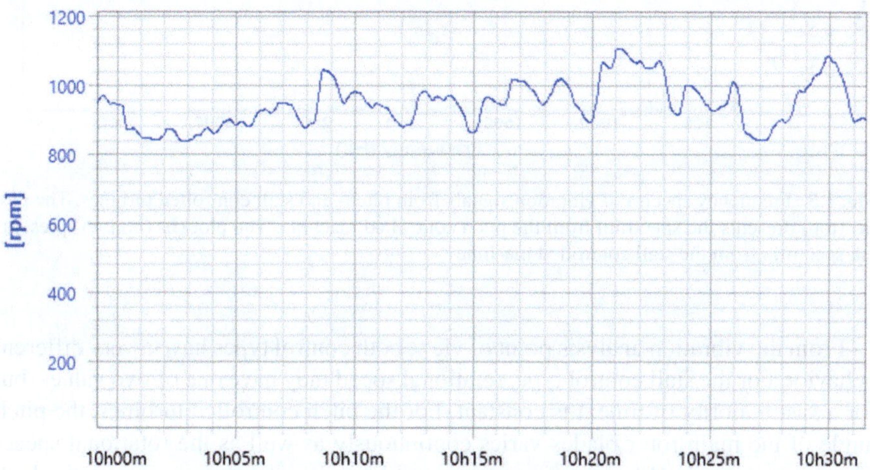

Fig. 1.4 Example of speed profile of a pitch controlled wind turbine during 30 min period

approximately 18% and causes all the characteristic frequencies to move on spectrum by this relative difference.

The additional benefit of a pitch control design is that it can be used as a braking system in extremely high wind conditions. According to the regulations, each turbine must be equipped in two independent braking systems, so pitch controlled turbines do not need any additional investment in yet another brake. Currently, all the high power wind turbines are designed and equipped with a pitch control.

The Fig. 1.5 presents comparison of spectra from these two types of turbines. The plot (a) presents the spectrum from the pitch controlled turbine. One can observe blurred frequency lines as a result of continuously varying rotational speed. The plot (b) presents the spectrum from the stall controlled machine in which the fixed speed of stall controlled turbine keeps all the spectral lines in a fixed position in the spectrum. The lines are sharp with no blurring effect. Several characteristic frequencies were identified and marked on the spectrum plot.

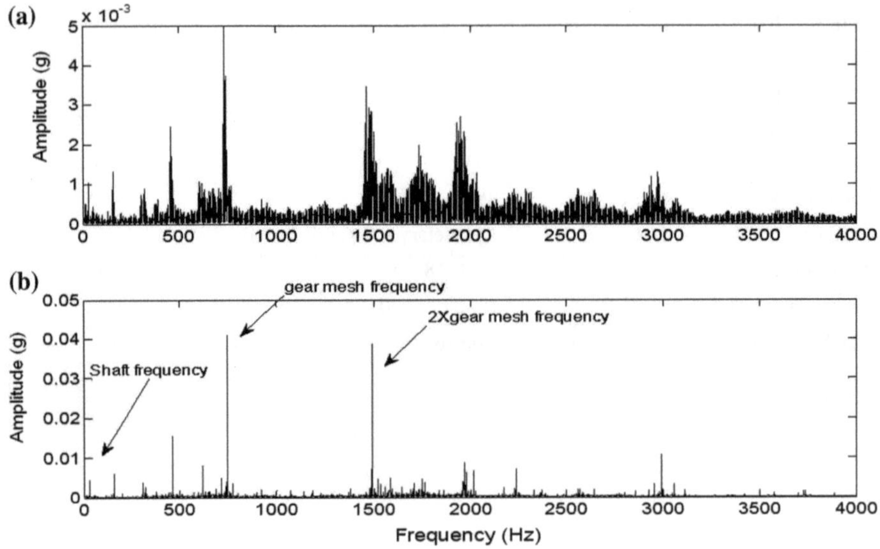

Fig. 1.5 Frequency spectra of vibration signals from pitch- and stall controlled turbines. The plot (**a**) (top) presents the spectrum from the pitch controlled machine. The plot (**b**) (bottom) presents the spectrum from the stall controlled machine

From the vibration analysis point of view, both control types have a very different behavior. For the stall control type, rotational speed may have one of two values, but for a vast majority of time it is constant. For the pitch controlled turbines, the pitch angle of the main rotor blades varies continuously as well as the rotational speed. These variations lead to smearing of spectral lines on a frequency spectrum which makes the detection and identification of frequency related features much harder. Vibration analysis of such signals requires more advanced techniques, called order analysis. They will be described in Chap. 2.

1.4 Key Machinery Parts and Their Characteristic Frequencies

Drivetrains of classic design wind turbines are complex machines whose role is to deliver torque of a rotor to a generator. Rotational speed must be increased by a factor close to 100. In order to accomplish this task, a multi stage gearboxes are used. The Fig. 1.6 presents the layout of a typical wind turbine drivetrain.

The gearbox design in the Fig. 1.6 consists of a single stage planetary gear and a two stage parallel gears. As it is shown, the drivetrain contains numerous shafts, gears and bearings. There are also other possible gearbox configurations, like a three stage parallel or a two stage planetary gears. The main task of condition monitoring

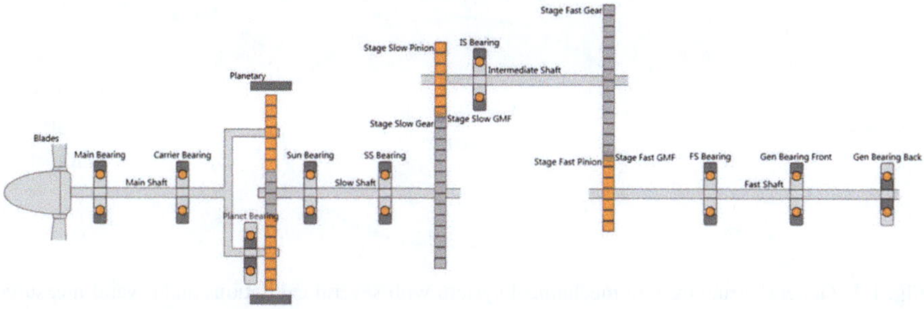

Fig. 1.6 Example layout of a wind turbine with a single planetary stage gear and a two stage parallel gear. Gears and bearings are depicted schematically and not to scale

is to assess technical state of the complete drivetrain and its individual parts. The key source of information is a vibration signal. Such a signal contains a mixture of many components.

The vibration based diagnostics uses signature analysis to distinguish between different sources of vibration signals originating from different machinery parts. From the vibration analysis point of view there are several groups of machinery parts:

- shaft,
- rotor with blades,
- coupling,
- gear,
- rolling element bearing.

Each of these groups represents a particular dynamic behavior and generates specific vibration. The most important parameter of each part is so called *characteristic frequency* which can be described as a repetition rate of a phenomenon bound to this part. All characteristic frequencies depend on rotational speed of a rotor. Thus, it is presented most easily as a ratio between the frequency and the rotational speed of the shaft. The following chapters will present only the most important characteristic frequencies for the key drivetrain components. The reader interested in more exhaustive information can refer to one of the several books on the subject of machinery diagnostics, e.g. works by Randall [3], Rao [4] or Klein [5].

1.4.1 Excitation Versus Structural Frequencies

All the aforementioned components may create a mechanical excitation at their respective characteristic frequencies. First, one should consider how the excitation can be observed in the vibration signal. In general, in any reasonably complex mechanical system there are several concurrent excitations and the system is

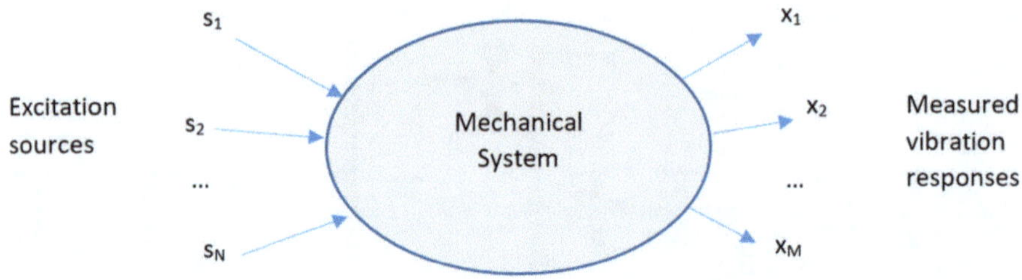

Fig. 1.7 General structure of a mechanical system with several excitations and several measured vibration responses

observed by means of vibration measurements in numerous points on the machine structure. Let us consider such a situation presented in a general way in the Fig. 1.7.

N excitations, named from s_1 to s_N, introduce energy into the mechanical system. The outputs, x_1 ... x_M, where in general case $N \neq M$, measure responses, which for each x is a mixed response from all the sources. Such a system is referred to as a multi input, multi output system (MIMO).

The structure presented in the Fig. 1.7 has important consequences. Firstly, there are many sources in the system and they all concurrently generate signals. The valuable diagnostic information is often related to a single particular source and the signal of interest (referred to as SOI), must be extracted using signal processing methods.

Secondly, all the excitation sources contribute to vibration in every point. Placing a sensor on one part of the machine will not prevent it from collecting a vibration signal from another part, though levels of vibration signals will certainly depend on the transfer path of a signal.

Thirdly, we never observe excitations directly, but only responses of the system. It is important to understand how this process can be described in a quantitative way. If we assume the linearity of the system, then response x_i caused by all the sources s can be described as:

$$x_i = \sum_j s_j * h_{ij} \tag{1.1}$$

where h_{ij} is the impulse response function between the source j and the output i.

Thus, the response is a sum of convolutions of all the sources and their corresponding impulse responses. The same operation can be simplified in the frequency domain and becomes a multiplication:

$$X_i = \sum_j S_j H_{ij} \tag{1.2}$$

where X and Y are, respectively, Fourier transforms of x and y.

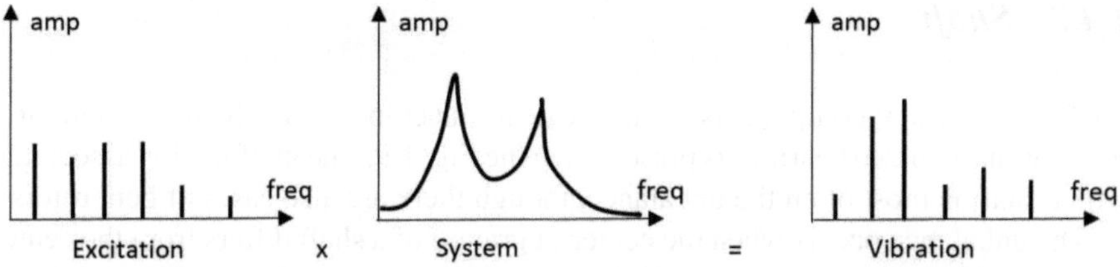

Fig. 1.8 Graphical presentation of discrete excitation and transfer function generating the discrete vibration output

In the Eq. (1.2) impulse transfer function becomes the Frequency Response Function (FRF) of the system. In the case of rotating machinery, the system is the structural response of the shafts, bearings, casings and all the components between the excitation source and the vibration sensor. Equation (1.2) is better presented on a frequency plot in the Fig. 1.8.

Since FRFs are complex, it is clearer to present them on a logarithmic scale. Logarithm of the FRF becomes amplitude which is a real value. Phase of FRF becomes the imaginary part and influences the phase of the response. Moreover, multiplication on the log scale becomes addition and it is easier to obtain the resulting vibration levels from the level of excitation and its corresponding structural response.

The Fig. 1.8 brings another important feature of machinery vibration. For machines running at a constant speed the majority of excitations have discrete frequencies, as these are well known repetitions caused by revolving motion. In a linear system the result is the same frequency with modified amplitude and phase. In reality, due to nonlinearities a single frequency excitation will generate the same frequency and a number of its harmonics at integer multiples of the fundamental excitation frequency. In real systems there are also broadband excitations caused e.g. by slip phenomenon in rolling bearings, wind turbulences etc. For illustration purposes, the example will be limited to discrete excitations only.

The structural response has a very different nature and is of a continuous essence. When excited by a broadband signal it will respond in a broad frequency range with peaks at frequency resonances. This difference helps to distinguish between excitation components and structural ones. It is important to distinguish whether a detected signal change is caused by a change of excitation (e.g. change of unbalance or misalignment) or a change in the structure (e.g. casing crack).

After this brief introduction to the process of vibration generation, various sources of excitations in wind turbines are going to be analyzed.

1.4.2 Shaft

Shafts are the key components of any rotating machinery. A schematic view of a rotating shaft on two bearings is presented in the Fig. 1.9. The malfunction associated with a shaft is most often the unbalance, though there are also cases of bent rotors.

The unbalance occurs when the center of gravity of a shaft differs from the center of rotation. The resulting force will depend on displacement between the center of gravity and the center of rotation, the mass of the rotor and the square of the rotational speed. The resulting vibration signature will consist of the rotational speed only. In other words, the frequency spectrum of vibration signal taken from an unbalanced shaft will show the peak at the frequency equal to that of the shaft. For example, when the shaft speed is 1500 rpm, we could see the increased spectral line at the frequency 25 Hz.

To facilitate the description of the rotational speed, the notation nX will be used in which n is a factor by which the rotational frequency should be multiplied to obtain the exact frequency. So, for the unbalance, the characteristic frequency is $1X$. In other words, if there is an unbalance on the disk rotating on a shaft we shall observe the once-per-revolution vibration signal.

Since the unbalance force is proportional to the squared rotational speed of a shaft, for wind turbines the force from the main rotor shaft will be very small, as the rotational speed is below 20 rpm. The Fig. 1.10 presents the main shaft and the main bearing of the wind turbine. The vibration sensor and the speed sensor are visible on the left.

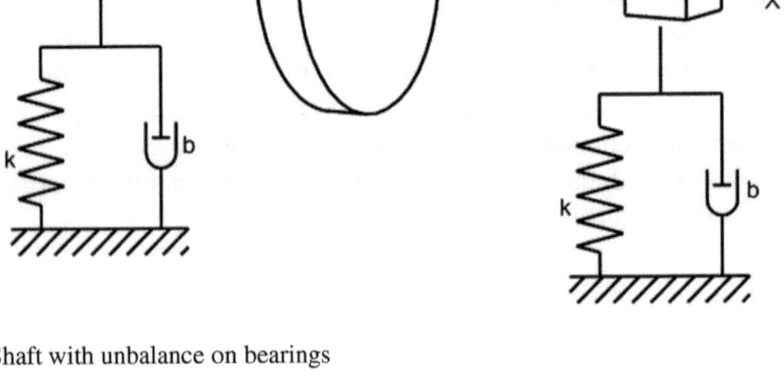

Fig. 1.9 Shaft with unbalance on bearings

Fig. 1.10 Shaft with main bearing (view from the gearbox side)

1.4.3 Bladed Rotor

The main phenomenon in a bladed rotor is passing of the blades over a stationary point in space. The characteristic frequency, called the blade pass frequency, equals:

$$f_b = N_b \cdot f_{shaft} \tag{1.3}$$

where:

f_b blade pass frequency [Hz]
N_b number of blades [–]
f_{shaft} rotational frequency of the main shaft [Hz].

Such a frequency is typical for fans, turbines, propeller pumps etc. In wind turbines such a behavior is observed in the main rotor. Most contemporary turbines have rotors with three blades. Thus, since the main rotor speed is ca. 18 rpm, the blade passing frequency is 54 rpm or about 1 Hz.

A wind turbine rotor is itself a complex system, especially when equipped with a pitch control mechanism. The Fig. 1.11 presents a part of the main rotor on a pitch control turbine. Parts of two blades are visible above the hub. The pitch control system may also be a subject of condition monitoring. Another important consideration is

Fig. 1.11 View from the nacelle on the main rotor. Part of the rotor hub and part of two blades are visible

the technical state of blades. Each one may be 30...40 meters long or even longer for multi–megawatt turbines. The main rotor blades may experience several faults, like delaminations, cracks or other material faults, but none of these are subject of this book.

1.4.4 Coupling

A coupling is used to connect a high speed shaft from the gearbox with a generator shaft. The malfunction often caused by the coupling is misalignment. There are three types of misalignments: parallel, angular and mixed. The Fig. 1.12 presents the scheme of a coupling with a mixed misalignment.

The exact manifestation differs depending on the type of a machine, but in general, misalignment increases the second harmonic of the vibration signal and to a lesser extent—the first harmonic. The characteristic frequency f_{mis} generated by a misaligned bearing is given by the Eq. (1.4):

$$f_{mis} = 2 \cdot f_{shaft} \tag{1.4}$$

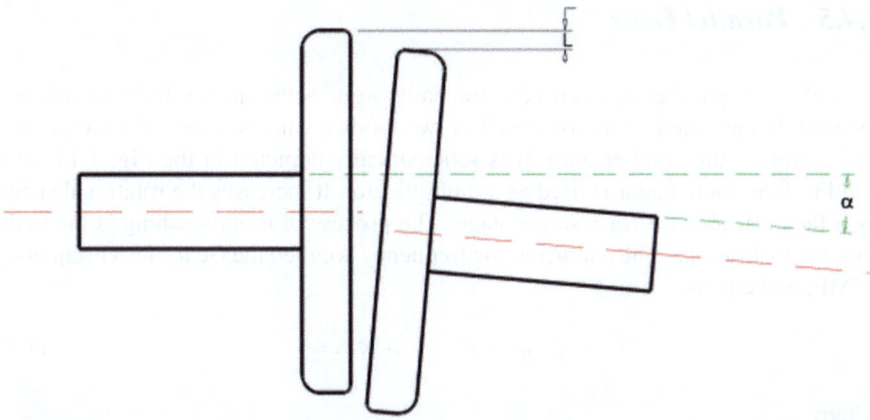

Fig. 1.12 Scheme of coupling with parallel and mixed misalignment

Fig. 1.13 Top view to the coupling between the gearbox (left) and the generator (right). The brake disk is visible on the left

A coupling is also used for measurement of the rotational speed. For condition monitoring purposes the 1-per-rev sensor is used (typically magnetic). Such a sensor can be spotted in the Fig. 1.13 on the mounting bracket on the right.

1.4.5 Parallel Gear

A single stage parallel gear is used to transmit torque between two shafts and change their rotational speeds. It is composed of two toothed wheels, a gear (the larger one) and a pinion (the smaller one). It is schematically depicted in the Fig. 1.14. In a wind turbine such a gear is used as a multiplicator. It increases the rotational speed by a factor of about 5 for a single stage. The process of tooth meshing is the main source of vibrations. The characteristic frequency is called the Gear Mesh Frequency (GMF) and equals:

$$f_{GMF} = N_1 \cdot f_1 = N_2 \cdot f_2 \tag{1.5}$$

where:

N_i number of teeth on a wheel i [–]
f_i rotational speed of that wheel i [Hz].

It is important to mention that the modulation phenomena shape the vibration signal from gear pairs. There are two fundamental types of modulation, namely the amplitude modulation (AM) and the frequency modulation (FM). They will be described in the Chap. 2. Apart from the fundamental GMF frequency, its harmonics are also present in the vibration signal as a result of amplitude modulation. Additionally to GMF harmonics, gearboxes also exhibit frequency modulations of the signal. The FM processes create families of sidebands around the center gearmesh frequency and its harmonics. Sidebands are additional spectral lines which are separated from the fundamental frequency by a constant frequency. In gearboxes this spacing frequency is determined by the rotational speed of the wheel. As there are

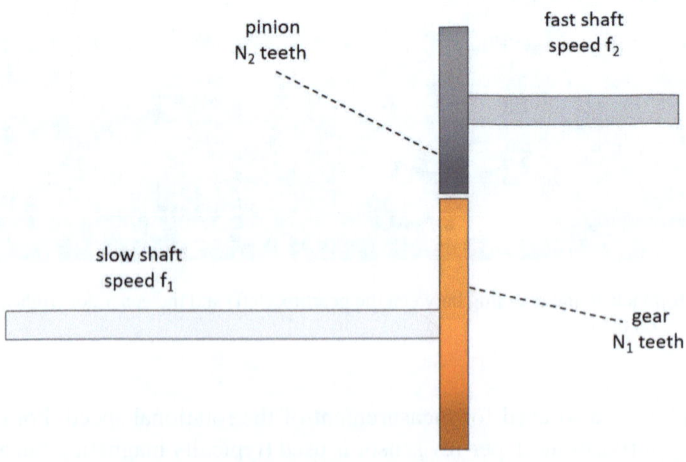

Fig. 1.14 Layout of a single stage parallel gear

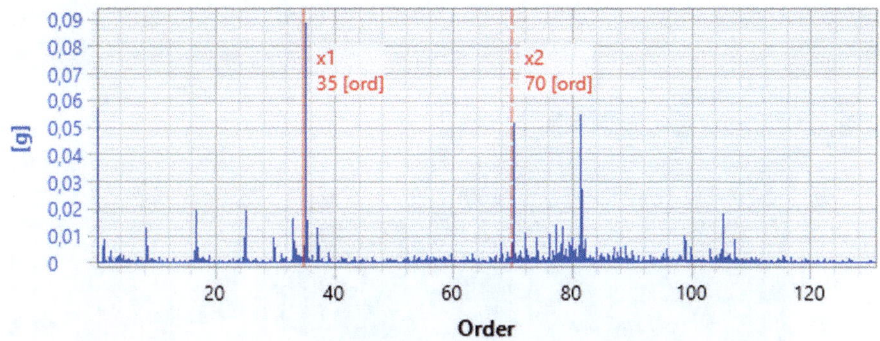

Fig. 1.15 Example of a vibration spectrum generated by a single stage gearbox. GMF and its second harmonic are marked by the cursors

two wheels, there are two families of sidebands, too. The Fig. 1.15 presents the frequency spectrum generated by a parallel gearbox.

The existence and amplitude of gear mesh frequencies and sidebands is a very good indicator of the gearbox health. The detailed presentation of gear vibration involves complex phenomena and was described e.g. by Randall [3]. In general, the process of meshing of a new gear is strictly periodic and generates vibration patterns with clear harmonics of GMF. Additionally, tooth deflection which depends on the load, introduces a frequency/phase modulation to the signal. In the vibration signal it can be observed as a series of sidebands around the GMF harmonics. The number and amplitudes of sidebands grow with the increasing wear. For the worn parallel gearbox being in a very bad condition, with looseness on bearings etc., one can hardly detect gear mesh frequencies and the energy of the whole frequency region around them has an increased level.

Gearboxes used in wind turbines most often combine gears of both types, i.e. parallel and planetary. The Fig. 1.16 presents a gearbox with a single planetary stage and two parallel stages. A vibration sensor installed in the horizontal direction is visible in the center.

1.4.6 Planetary Gear

Planetary gears (also known as epicyclic gears) have a more complex design with three distinct components, i.e. a ring, a planet carrier with satellites and a sun. Planetary gears are more expensive to manufacture, but they can achieve higher ratios (about 10) and can transmit higher loads. In theory, torque may be transmitted from any of these components to another one with the third being stationary. In planetary gearboxes of a wind turbine a ring is stationary and torque is transmitted from a planet carrier to a sun, as depicted in the Fig. 1.17.

In such a configuration, the ratio of the gearbox n is given by the formula:

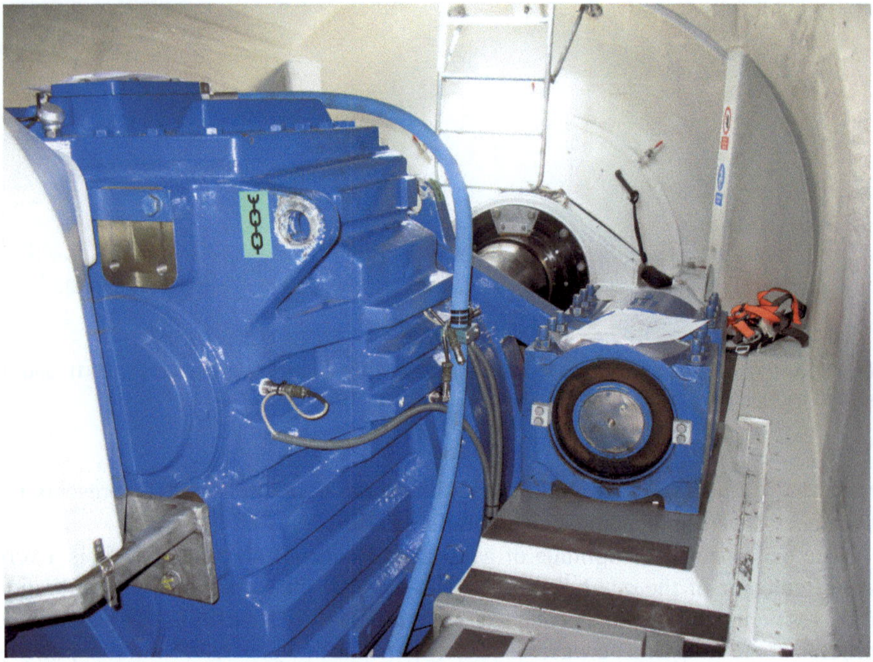

Fig. 1.16 Right side of the gearbox (view in the direction of the main rotor)

$$n = \frac{N_s + N_r}{N_s} = 1 + \frac{N_r}{N_s} \tag{1.6}$$

Or, in other words, the rotational speeds of planet carrier and sun are related to each other as:

$$f_s = n * f_c = \left(1 + \frac{N_r}{N_s}\right) f_c \tag{1.7}$$

Following symbols are used for the planetary gear frequencies:

N_s number of sun gear teeth
N_p number of planet gear teeth
N_r number of ring gear teeth
n_p number of planets
f_c rotational speed of the planet carrier
f_s rotational speed of the sun
n gear ratio.

The vibration pattern of planetary gears is similar to that of parallel gears, as the main source of excitation is the meshing phenomenon. In any fault involving a gear, the GMF will be excited and visible on the spectrum, together with its harmonics.

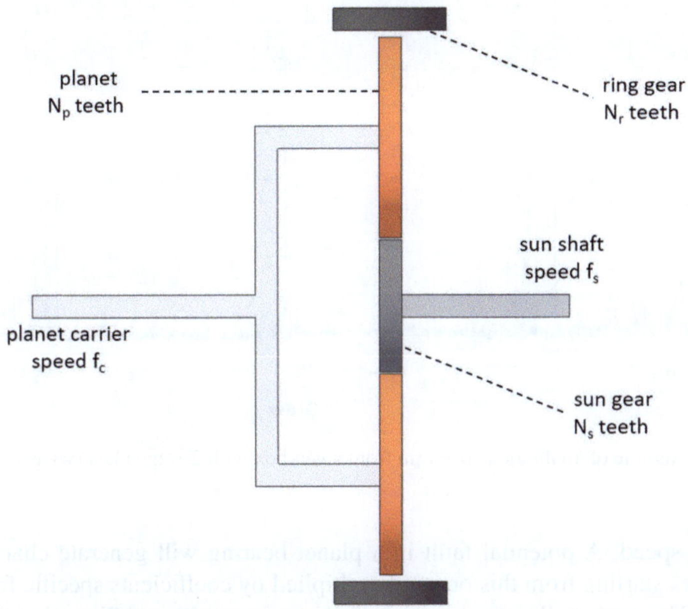

Fig. 1.17 Layout of a single stage planetary gear in the configuration used in wind turbines. The ring is stationary, input is the planet carrier, output is the sun shaft

Table 1.4 Characteristic frequencies of the planetary gear faults

Description	Characteristic frequency	Location of fault
Gear Mesh Frequency (GMF)	$N_r \cdot f_c$	Present in majority of failures
RingOverRoll	$n_p \cdot f_c$	Ring gear
SunOverRoll	$\frac{N_r \cdot n_p}{N_s} f_c$	Sun gear
Planet Overroll	$\frac{N_r}{N_p} f_c$	Planet gear
Planet REB	$\frac{N_r - 2N_p}{N_p} f_c$	Bearing of the planet

However, apart from the GMF a planetary gear exhibits more periodicity in its operation. Therefore, the number of characteristic frequencies is higher than for a parallel gear. If there is a local fault on a parallel gear part, it will periodically come into contact with a gear from another part and excite a modified pattern, most often in a form an impulse. Spacing of these impulses can reveal the location of the fault. Each part of a gear, i.e. a ring, a planet and a sun has a different period of coming in contact with other parts. These periods can be converted into frequencies and used for frequency analysis. The Table 1.4 presents the fault frequencies for a planetary gear.

The last row presents the frequency of the rotational speed of the rolling element bearing on a planet. It is worth mentioning that these bearings have their own

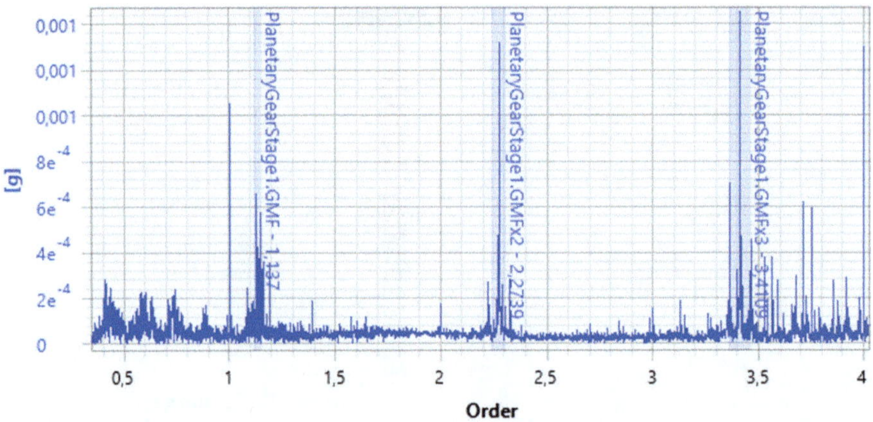

Fig. 1.18 Example of a vibration spectrum from a gearbox with 2-stage planetary gears

rotational speed. A potential fault in a planet bearing will generate characteristic frequencies starting from this one and multiplied by coefficients specific for rolling bearings. This topic will be further presented in the next chapter. Planet bearing faults are hard to detect due to the complex and varying transfer path.

Please note, that in general epicyclic gears are quite complex and can be designed in several variants. The formulas from the Table 1.4 are given with the assumption of a stationary ring and torque being transmitted from a planet carrier to a sun.

In Fig. 1.18 the example of a vibration spectrum from the planetary gearbox is presented. On the spectrum one can observe the GMF with its second and third harmonics. The sidebands around each GMF harmonic are visible as well.

1.4.7 Rolling Element Bearing

Rolling element bearings (REBs) are components which transfer load through elements in rolling contact. There are several types of these bearings depending of a rolling element type (ball, roller, spherical etc.), but all of them share the same principle of operation. The Fig. 1.19 presents the layout of a rolling bearing. The REB consists of: an inner race, an outer race, rolling elements and a cage which holds the rolling elements in a given relative position. Detection of REB component faults is very important, as it prevents any further deterioration to other components which may lead to a catastrophic failure. A common failure case starts with deterioration of a rolling bearing. When the level of the fault is large enough, wear products are transported with oil to other components, e.g. gears where much larger damage is done.

Rolling element bearings fail in a number of different ways (spalling, brinnelling, etc.) and as a result of a number of factors (fatigue, oil contamination, overload,

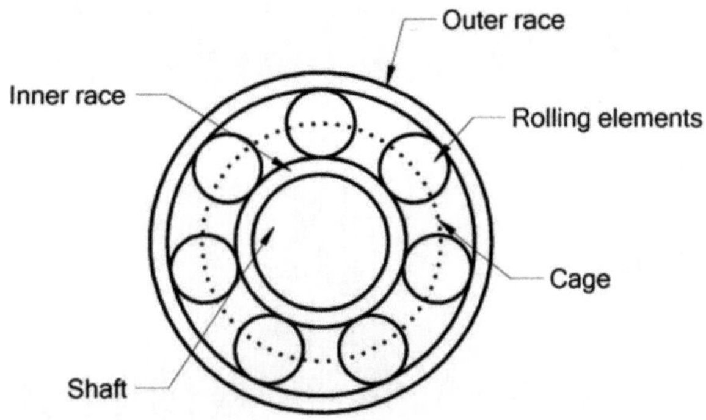

Fig. 1.19 Rolling element bearing components

Table 1.5 Characteristic frequencies of rolling element bearing faults

Fault description	Characteristic frequency	Location of fault
Ball Pass Frequency of the Outer race (BPFO)	$f_r \frac{N_r}{2}\left(1 - \frac{D_b \cos\varphi}{D_p}\right)$	Outer race
Ball Pass Frequency of the Inner race (BPFI)	$f_r \frac{N_r}{2}\left(1 + \frac{D_b \cos\varphi}{D_p}\right)$	Inner race
Ball Spin Frequency (BSF)	$f_r \frac{D_p}{D_b}\left(1 - \frac{D_b^2 \cos^2\varphi}{D_p^2}\right)$	Rolling element
Fundamental Train Frequency (FTF)	$f_r \frac{1}{2}\left(1 - \frac{D_b \cos\varphi}{D_p}\right)$	Cage

etc.). Fatigue results in the appearance of spalls on the inner race, outer race or rolling elements.

Behavior of REB fault at an early stage (so called localized fault) follows a typical pattern. If one of the races has a spall, it periodically impacts on rolling elements. The fault signature is represented by a series of impulses with a repetition rate depending on a faulty component, geometric dimensions and the rotational speed. The period between impacts is different for all the listed elements and depends on geometry of a bearing, the rotational speed and the load angle. It also depends on which race is stationary and which one is rotating. For a fixed outer race bearing, as it is the case in wind turbines, the theoretical ball bass frequencies are given in Table 1.5 [5].

where,

f_r Rotational speed of the shaft (Hz)
D_b Rolling element diameter
D_p Pitch diameter
N_r Number of the rolling elements (for a single row)
φ Load angle (contact angle from radial).

A vibration pattern of a faulty rolling element of a bearing is a combination of two phenomena. As mentioned before, it is a series of impulses. Each impulse is an excited transfer path between a bearing race and a vibration transducer. It is

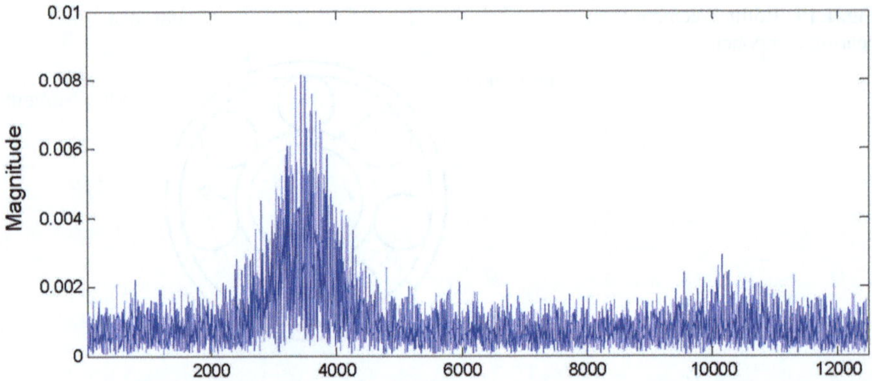

Fig. 1.20 Example of a vibration spectrum from a series of impulses exciting a structural resonance

in general a complex system, but it results in excitation of a broadband structural resonance. As mentioned in Sect. 1.4.1 on the frequency spectrum it looks like a broad structure. The actual frequency range greatly differs depending on a machine design and material used, but typical ranges are between 5 and 10 kHz. Higher frequencies may also be excited, but that band is used by ultrasonic methods. The Fig. 1.20 presents a spectrum of a series of impulses exciting a resonance which is a representative for a typical faulty REB. In a real case, however, a vibration signal also contains several discrete excitations from shafts, couplings, gears and others which mask the resonance caused by a REB fault.

An important fact about the detection of REB faults is that we are not interested in excited frequencies themselves, but in a repetition rate with which they are excited. A repetition rate is very small compared to structural resonance frequencies (a few Hz to a few hundred Hz compared to a few kHz). A repetition rate is often not visible on a frequency spectrum at all. Therefore, the most efficient method to detect REB faults is the envelope analysis, which effectively demodulates (i.e. shifts) a vibration signal, discards structural resonance frequencies and shifts a repetition rate to low frequency range. The envelope analysis will be described in the next chapter and followed by a few examples.

1.4.8 Wind Turbine Characteristic Frequencies

A wind turbine drivetrain under operation generates a complex vibration pattern. Its most important components are excitations caused by rotating elements (shafts, couplings, gears, etc.) and structural responses from other components (casings, pedestals etc.) which modify the original vibration signals. Since structural resonances do not change and sensor locations are constant, vibration monitoring is focused on excitation frequencies. These frequencies are often referred to as charac-

Table 1.6 Example kinematic data of the 1 MW stall-controlled wind turbine. All the values in the 'Relative frequency' column are related to the generator shaft frequency (reprinted from (Barszcz and Randall 2009) with permission from Elsevier)

Parameter	Relative frequency	Frequency at 1500 rpm [Hz]	Period at 1500 rpm [ms]
Generator shaft	1.0000	25.0	40.0
Intermediate shaft	0.3256	8.14	122.8
Sun shaft	0.0754	1.885	530.5
Planets	0.0294	0.735	1360
Plate (main rotor)	0.0119	0.2975	3361
Meshing parallel gear II	28.000	700	1.429
Meshing parallel gear I	7.1628	179.07	5.584
Meshing planetary gear	1.206	30.15	33.17
Overroll ring	0.0358	0.895	1117
Overroll planets	0.0294	0.735	1360
Overroll sun	0.1904	4.760	2100

teristic frequencies. There are many of such frequencies in a wind turbine. Including rolling bearings they can exceed 100 per machine. Table 1.6 presents the example of basic characteristic frequencies of a wind turbine. This turbine is stall controlled with a gearbox with one planetary stage and two parallel stages. In order to be able to compare the frequencies, it was rescaled to that of the generator shaft which is the highest speed shaft.

One can see large span of frequencies in the drivetrain, the lowest being the main shaft (0.0119) and the highest—GMF of the second stage of the planetary gear (28.0). If four first harmonics are taken into account, the frequency span is $112.0 : 0.0119 = 9411$, which is almost 4 orders of magnitude. It is an unusually wide range, requiring high performance signal acquisition and processing system.

1.5 Varying Operational Conditions

The varying operational conditions (VOC) are the key factor influencing vibrations of several heavy duty rotating machines including wind turbines. Due to the inherent variability of the wind, both frequency and amplitude of vibration components change. The Fig. 1.21 presents a typical pattern of variability of wind speed, generator output power and rotational speed of a generator shaft over a period of 60 min.

The input variable of a wind turbine system is the wind speed. In response to it a controller sets rotational speed and a turbine responds in the output power value. The time resolution of measurements is 1 s, so the dynamic behavior can be tracked as well. A wind signal has the highest amount of high frequency components and consequently, the richest spectral content. Both rotational speed and power are cor-

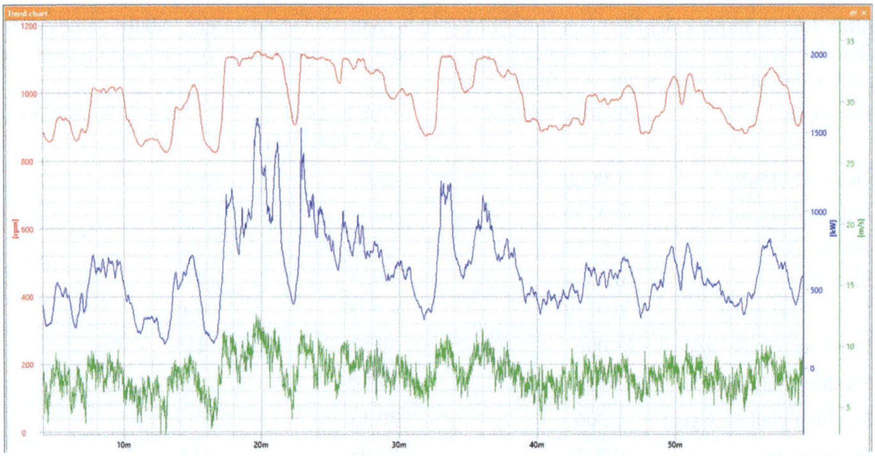

Fig. 1.21 Pattern of variability of wind speed (bottom, green), generator output power (middle, blue) and rotational speed of a generator shaft (top, red). The plot covers a time period of 60 min

related to the wind speed, but can be treated as the output of a low pass filter. Still, these two signals also exhibit great variability. During the presented period all the parameters vary constantly, but there are events when wind speed can increase from 3 m/s to 11 m/s in just 60 s (minute 16–17). The relevant change of the output power is 150 to 1200 kW which is a very significant change from almost an idle condition to the nominal power.

As shown by Barszcz et al. [6], wind can be described as a chaotic process which can be modelled with a fractal approach. The irregular character of the wind speed was investigated over a set of 300 samples. The Weierstrass function was used for modelling of the wind speed. The interesting feature of this function is that it describes a fractal set which is a good choice for modelling wind speed as a time series. It seems that the irregular character of the graph of Weierstrass function caused by the fact that in each point it is continuous but nondifferentiable, can reflect chaotic irregularities of wind speed as a time series correctly. This assumption was confirmed during the research. Actually, it confirmed the original suggestion by Richardson [7] from as long ago as 1926, but was never proven before. Further research presented in Bielecki et al. [8] showed how this variability can be used to model a stochastic load of a wind turbine.

Such a rapid change of operational conditions certainly has an impact on the dynamic state making the use of standard vibration analysis techniques harder to apply. The Fig. 1.22 presents the example of how vibration can change due to the change of operating conditions. The operating conditions are: wind speed (green), generator rotational speed (red) and generator output power (black).

The vibration signal has the length of 110 s. During this time the vibration level changes by more than 40% only due to the change of operating conditions with no change of the technical state. The correlation between the vibration amplitude and all

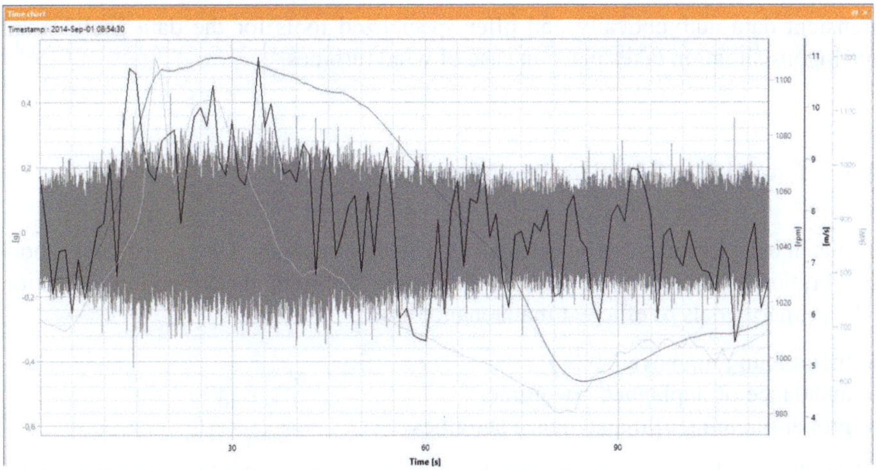

Fig. 1.22 Example of a change of a raw vibration signal due to the change of wind turbine operating conditions. The operating conditions are: wind speed (black), generator rotational speed (dark grey) and generator output power (light grey)

the process variables can be clearly seen. Changes caused by operating conditions are much larger than these caused by faults of the mechanical components themselves. It is necessary to take these variability into account and separate one process form the other.

The highly variable operation has a large influence on wind turbines. The load is varying rapidly, changing from no load to overload. This, in turn, causes sudden acceleration and deceleration of a drivetrain. Such a variability causes a subsequent different dynamic behavior of gears. The most important consequence is the decrease of longevity of wind turbine drivetrains. There are a number of cases in which many wind turbines with a supposed design life of 20 years are failing after only five years (Barszcz and Randall 2009). It happens when heavy duty gearboxes from other industrial applications are used in wind turbines. Here they are exposed to very different load patterns of a much higher variability. Fatigue processes are much faster and they result in a decreased lifetime and premature failures.

From the signal processing point of view, variable operating conditions are the cause of varying of the vibration signal spectral content. It also calls for a fact that diagnostic features should be dependent on a load and other process variables. Therefore, condition monitoring methods for wind turbines should be different from those for the other machinery operating in more stable conditions.

It is worth mentioning that there are also other types of machinery which require a special treatment of data. For example, large steam and gas turbines have journal bearings which exhibit nonlinear behavior. These bearings are of vital importance for the safety of the turbine operation. Since the most valuable data are taken during a run-up and a coast-down of these machines, there are special condition monitoring systems suited for this type of operation. They are capable of collecting so called transient data automatically and offer specialized tools for the data analysis. An analogous situation takes place in case of wind turbines.

1.6 Condition Monitoring Methods

Faults in wind turbines can be detected by different methods. The subject of this book is vibration analysis, but the other methods will be briefly described in this section. The primary goals of efficient maintenance are:

- maintaining high availability,
- avoidance of unplanned downtime,
- proper planning of maintenance activities.

High availability is directly correlated to the return on the investment in a wind farm. Higher availability means more of the wind resource being converted into electrical energy and thus, into a profit for a farm owner and all stakeholders. For maintaining high availability and controlling costs of operation it is of utmost importance to reduce unplanned downtime. Such downtime requires organizing repair actions, usually on a very short notice. This, in turn, affects costs of manpower and spare parts. The specific factor of the wind power generation is a high cost of maintenance. If a large component is to be serviced or replaced, it requires costly equipment, e.g. high cranes to be transported to the site. Such operations can only be performed when weather conditions are permitting. There are, for example, limits on wind speed when a crane can be operated. Conditions are even tighter for offshore farms where transport often requires specialized ships and later, the sea level also needs to be taken into account. Sometimes, if a fault takes place, the owner decides to wait until the next scheduled maintenance is to take place and use this time slot for the repair. It limits extraordinary costs of repair, but reduces the income due to an idle state of the faulty turbine. Proper planning of maintenance is of vital importance for a profitable operation of wind farms.

Taking decisions about maintenance activities requires information about the actual state of turbine components and possibly an accurate prediction about the evolution of this state until the next maintenance slot. Such information can be obtained by a number of sources and techniques. According to MarketsandMarkets survey [9], there are several techniques used, like vibration monitoring, thermography, ultrasound emission, lubricating oil analysis, corrosion monitoring and motor current signature analysis.

Another important distinction, apart from the technology, is the monitoring mode. Diagnostic techniques may come in solutions suited for online or offline monitoring. Online monitoring is more common for modern wind turbines and provides constant stream of the technical condition data about the monitored turbine. The CMS operates continuously, so even intermittent faults can be detected and later analyzed. On the other hand, the online system requires high investment, as it needs to be purchased, installed and configured for each turbine separately.

The other solution is an offline system. Such a system is portable and can be placed on a machine for a chosen period of time, from as short as a few minutes to several weeks (though not all portable systems are suitable for such long monitoring). Data from a portable system deliver information about the drivetrain and are analyzed later in the office. The application of an offline system requires much lower investment. On the other hand, some faults may go unnoticed, for example if a fault depends on specific environmental conditions. Using offline systems requires more human workload, as the system needs to be transported, moved to the nacelle, and data need to be transferred and analyzed.

1.6.1 Vibration Analysis

Vibration analysis is by far the most important method used for condition monitoring. According to the market survey mentioned above, vibration analysis alone constitutes about 58% of the total CM market.

Vibration based monitoring is the most often used method as it uses accelerometers which are relatively low cost sensors. This method can provide good tracking of the development of a fault. Another important feature is the ability to pinpoint the location of a fault. Faults at an early stage are localized, and so each kinematic pair in a rotating motion periodically comes into contact and generates a so called signal of interest (SOI) or a fault signature. For most pairs in a drivetrain these periods are different which gives a possibility to use the information to extract the SOI and use it for the exact localization of a fault.

Vibration based condition monitoring methods are the subject of this book and will be discussed in detail in the next chapters. As the introduction to the subject, numerous sources can be advised for the interested reader. It is a rapidly developing field and during the past decades a number of vibration signal processing techniques have been established. Comprehensive surveys of such techniques are available and can be found in the classical books by Randall [10] or Klein [5]. The very classic and rich review was presented by Jardine et al. [11]. Another, a more recent review, was presented by García Márquez et al. [12]. Several reviews focused on vibration based wind turbine monitoring can be found, but they are relatively narrow, e.g. Nie and Wang [13] or describe only a given test rig research, like the Siegel et al. [14].

1.6.2 Ultrasonic Analysis

The ultrasonic analysis is an extension of the most basic condition monitoring method for rotating machinery which is a human ear. Hearing is the sense particularly useful for analysis of rotating motion and we all use it in our everyday life. In a way, the development of CM methods is a history of attempts to develop an automated and objective "artificial ear". Even when these methods are used for mechanical vibration, data acquisition and used analysis methods share a significant amount of expertise with acoustic analysis. Yet, it is a separate part of the market and constitutes about 8% of the CM market.

Ultrasonic analysis is an extension of the standard acoustic analysis and takes advantage of the fact that in many physical phenomena short impacts are generated, which—in turn—propagate through metal or air. Ultrasound waves experience less external noise than acoustic ones. Typical applications of ultrasonic analysis is the detection of incorrect lubrication of bearings (both under- and over-lubrication), early signs of wear of gearboxes and motors, leaks and electric discharges. The first two are important for wind turbines. The last one may be used for quick testing of electric equipment.

Monitoring of high frequency ultrasonic waves is sometimes referred to as the acoustic emission (AE) analysis. There are interesting works in which this technique is used for fault detection of rolling bearings. A significant advantage of this method is the ability to detect faults in slowly rotating equipment, e.g. Aye et al. [15], which is the case for a main bearing and a planetary gear in wind turbines.

The disadvantage of the acoustic emission method is its cost. It requires costly transducers and signal conditioners which are more expensive than vibration based systems. Nevertheless, it is an interesting direction of research. The interested reader may refer to works of Mba [16] or He et al. [17].

1.6.3 Oil Analysis

Friction is the main physical phenomenon responsible for creating wear in mechanical systems. In order to limit the wear, one must avoid dry friction by application of different types of oil or grease. Then, in normal conditions, machinery operates in the area of fluid lubrication (in which a lubricant layer is much thicker than inaccuracies of the surface). The boundary lubrication, when the layer of lubricant is comparable to the inaccuracies of a surface, must be avoided. Apart from avoiding friction, an important function of lubricant is corrosion protection and cooling.

Lubricants in machines are subject to wear which is mainly influenced by: pressure between mating teeth, speed of slip and temperature. The most popular approach towards oil maintenance is its exchange, according to the manufacturer's recommendations. This is widely known as preventive maintenance and it has become very popular. In a growing number of wind turbines, however, the owners implement predictive maintenance also referring to oil maintenance. Such maintenance requires active approach and not just a simple exchange after a given number of hours of operation. One can say that it is a change in perception, as oil is treated as an important component of a machine. It is so, because oil gets in contact with every other machine component, it is subject to physical and chemical interactions and it transports contamination originating from the contact area in gears and bearings (but also external ones).

One can say, that oil analysis has much in common with performing a blood analysis test on a human. Degraded lubricant symptoms are good indicators of equipment degradation. The analysis makes it possible to track the decrease of oil effectiveness. Using this tool one can estimate the approaching end of oil lifespan as well as detect any contaminants. Particles in oil are very dangerous as they may cause rapid wear and tear of components. Such debris generated by a fault in—let's say a bearing—are transported with oil throughout the whole machine and further may damage other components, like gears. A significant disadvantage of this method is its inability to locate the fault, as there is a single sensor in the oil pipe which simply measures oil parameters. On the other hand, it is a very efficient way to detect early signs of wear and tear of a turbine. This type of analysis has gained attention of researchers, as it is a very efficient supplement to classical vibration monitoring. The interested reader can refer to the paper of Loutas et al. [18] in which a combination of vibration, acoustic emission and oil monitoring was proposed.

There are several online sensors which are capable of monitoring oil quality in a machine. First of all, there are so called particle counter sensors, which can detect and count particles in oil. Such sensors can also estimate the size of a particle and calculate separately a number of particles in a few bins representing different sizes. Other oil sensors can monitor contamination, e.g. the amount of water in oil.

There are also offline solutions which are used for oil monitoring in the industry. Such offline analysis can be performed by sending the oil sample to a specialized laboratory or there are portable automated lab systems, capable of performing such an analysis in the field. The method used in such labs is spectrographic analysis. It can determine the elemental composition of oil. With the knowledge of chemical elements in oil, one can determine the source of contamination or the location of the accelerated wear. The size of debris is also determined in such a test.

Oil analysis is offered by numerous companies across the world. This service is used by many industries including wind power generation. It is an established part of the market and oil analyses constitute 11% of the market, according to the study cited above.

1.6.4 Electrical Parameters Analysis

A wind turbine comprises a mechanical system which converts kinetic energy of wind into mechanical energy of a generator shaft. An electrical system—in turn—converts mechanical energy into electrical energy. The electrical system consists of a generator, electrical auxiliary systems and a control and supervisory system. Faults in the electrical system do occur, but do not cause such long outages nor high financial losses. For proper condition monitoring and maintenance planning the electrical system can also be monitored with both offline and online techniques. Analysis of electrical parameters is much rarely used than other methods and accounts for 2% of the CM market only.

A prominent method for electrical machinery monitoring is the Motor Current Signature Analysis (MCSA). The method uses high frequency acquisition of voltage and current signals. After the frequency analysis, it can not only be used for detection of faults in electric components, but also in mechanical ones. The commercial CM systems are used for a broader range of monitoring and fault detection in electric machinery. These methods can be used for detection of faults of vital components of electric machines, like: a power circuit, a rotor, a stator, insulation, an air gap and power quality. The interested reader can find the detailed description in the book by Benbouzid [19] or in the recent one by Thomson and Culbert [20].

Electrical analysis can be used in two approaches, de-energized and energized. The de-energized CM systems require that an electric machine is detached from the grid and power supply. The energy is supplied from a tester. Only in such a configuration it is possible to test the insulation of the generator and auxiliary systems. The energized systems are connected to the running generator using 3 voltage probes and 3 current probes. It is capable of performing the MCSA analysis and assess other aspects of the operation of an electrical machine, e.g. power quality.

1.6.5 SCADA Data Analysis

The application of various condition monitoring systems is growing in the wind power generation industry. A significant obstacle to this trend is a need to invest in every CMS, with typical cost being in the range of at least a few thousand of Euro per turbine. For many turbines, this cost is regarded as too high and the CMS is not installed. On the other hand, every wind turbine is equipped with the Supervisory Control And Data Acquisition (SCADA) system which collects a number of data. These data include among many others: wind speed and direction, ambient temperature, power output incl. power quality measurements, temperatures of vital WT components (a gearbox, oil, generator winding, cooling water etc.) and control settings, like setpoint and actual: pitch angles, shaft speed and output power. Hundreds of analogue signals are collected. They are accompanied by binary signals informing about the position of dozens of switches and sequences of error codes that carry information about the status of many subsystems of a turbine. The data storage rate is relatively low and is typically 10 min. For some systems, every 10 min not only the actual value is stored, but also a minimum, maximum and standard deviation for the period.

This data stream is available for probably all commercially installed wind turbines and for a long time has been regarded as a valuable resource to determine the technical condition and prognosis of the evolution of the state. To analyze the data a number of methods have been proposed. The interested reader can refer to a review by Tautz-Weinert and Watson [21]. The methods range from a simple trending which can be very useful but requires significant human assistance to more advanced methods taking into account several variables and using a clustering approach to determine the operational condition of the machine.

Numerous Artificial Intelligence tools have also been developed. When used to detect a fault and follow the supervised learning scheme, they all face the problem of a relatively small pool of available faulty data. This is especially striking when compared with an enormous number of fault scenarios. It results in only very small coverage of the data from faulty states. Another popular approach is novelty detection, also referred to as anomaly detection. This model is less aspiring and aims at assisting the operator in finding new (or anomaly) data, which may mean a fault. For such systems the tool uses historical data, when no fault was recorded and uses it to train the AI system. These methods are not the subject of this book. Apart from the mentioned survey, the interested reader can find a comparison of commercial tools using the SCADA data in the survey by Chen et al. [22].

References

1. Hau E (2013) Wind turbines. Fundamentals, technologies, application, economic. Springer-Verlag Berlin Heidelberg
2. Burton T, Jenkins N, Sharpe D, Bossanyi E (2011) Wind energy handbook, 2nd edn
3. Randall RB (2011) Vibration-based condition monitoring. Industrial, aerospace and automotive applications. Wiley, Chichester
4. Rao JS (1996) Handbook of condition monitoring. Elsevier, Oxford
5. Klein U (2003) Schwingungsdiagnostische Beurteilung von Maschinen und Anlagen (Vibrodiagnostic assessment of machines and devices). Stahleisen Verlag, Duesseldorf (in German)
6. Barszcz T, Bielecka M, Bielecki A, Wójcik M (2012) Wind speed modelling using Weierstrass function fitted by a genetic algorithm. J Wind Eng Ind Aerod 109:68–78
7. Richardson LF (1926) Atmospheric diffusion shown on a distance-neighbour graph. Proc R Soc Lond A 110:730–737
8. Bielecki A, Barszcz T, Wójcik M (2015) Modelling of a chaotic load of wind turbines drivetrain. Mech Syst Sig Process 54–55:491–505
9. MarketsandMarkets Research Private Ltd. (2018) Machine condition monitoring market, by product (Vibration Monitoring, Thermography, Ultrasound Emission, Lubricating Oil Analysis, Corrosion Monitoring, and Motor Current Signature Analysis), component, application, and geography—Global forecast to 2024
10. Randall RB (1987) Frequency analysis. Bruel & Kjaer, Naerum
11. Jardine AK, Lin D, Banjevic D (2006) A review on machinery diagnostics and prognostics implementing condition based maintenance. Mech Syst Sig Process 20(7):1483–1510
12. García Márquez FP, Tobias AM, Pinar Pérez JM, Papaelias M (2012) Condition monitoring of wind turbines: techniques and methods. Renew Energy 46:169–178
13. Nie M, Wang L (2013) Review of condition monitoring and fault diagnosis technologies for wind turbine gearbox. Procedia CIRP 11:287–290
14. Siegel D, Zhao W, Lapira E, Abuali M, Lee J (2014) A comparative study on vibration-based condition monitoring algorithms for wind turbine drive trains. Wind Energ 17:695–714
15. Aye SA, Heyns PS, Thiart CJH (2016) Diagnostics of slow rotating bearings using a novel DAI based on acoustic emission. In: Chaari F, Zimroz R, Bartelmus W, Haddar M (eds) Advances in condition monitoring of machinery in non-stationary operations. CMMNO 2014. Applied condition monitoring, vol 4. Springer, Cham
16. Mba D (2003) Acoustic emissions and monitoring bearing health. Tribol Trans 46(3):447–451
17. He Y, Friswell MI, Zhang X (2009) Defect diagnosis for rolling element bearings using acoustic emission. J Vibr Acoust 131(6):1–10

18. Loutas TH, Roulias D, Pauly E, Kostopoulos V (2011) The combined use of vibration, acoustic emission and oil debris on-line monitoring towards a more effective condition monitoring of rotating machinery. Mech Syst Sig Process 25:1339–1352
19. Benbouzid MEH (1999) Induction motors faults detection and localization using stator current advanced signal processing techniques. IEEE Trans Power Electron 14:14–22
20. Thomson WT, Culbert I (2017) Current signature analysis for condition monitoring of cage induction motors: industrial application and case histories. Wiley-IEEE Press
21. Tautz-Weinert J, Watson SJ (2017) Using SCADA data for wind turbine condition monitoring—a review. IET Renew Power Gen 11(4):382–394
22. Chen B, Zappalá D, Crabtree CJ, Tavner PJ (2014) Survey of commercially available SCADA data analysis tools for wind turbine health monitoring. Durham University
23. Barszcz T, Randall RB (2009) Application of spectral kurtosis for detection of a tooth crack in the planetary gear of a wind turbine. Mechanical Systems and Signal Processing 23 (4):1352–1365

Chapter 2
Standard Vibration Analysis Methods

2.1 Background

Vibration signal processing is a very vast field of engineering knowledge and only popular methods used in engineering practice will be presented here. For an exhaustive course on signal processing, the interested reader can refer to several handbooks on the subject. The classic one by Randall [1] can be recommended as a good start, though it may no longer be available. More recent books are Shin and Hammond [2] with a very clear theoretical explanation of the fundamentals or Braun [3] where an example based approach was taken. Another book suited for engineers (who can speak German) is Klein [4]. Several review papers are also available to build the understanding of recent developments in the subject domain. A classic review was presented by Jardine et al. [5]. Another one, by Siegel et al. [6], presents a comparison of vibration based methods with focus on a wind turbine gearbox.

To facilitate the presentation of methods, the chapter will introduce a few signal types, from simple simulated ones to more advanced ones. Each consecutive step of simulation will bear more resemblance to real vibration signals. Signals from real wind turbines will be introduced as well. The presented methods will start with the analysis of a signal waveform itself. The other methods will be:

- frequency spectrum,
- envelope analysis,
- order analysis,
- time synchronous averaging,
- time-frequency transforms.

Based on these processing methods from a single vibration signal one can obtain numerous signal features. In general, they can all be divided into broadband and narrowband ones. They will also be presented in the later part of the chapter.

© Springer Nature Switzerland AG 2019
T. Barszcz, *Vibration-Based Condition Monitoring of Wind Turbines*, Applied Condition Monitoring 14, https://doi.org/10.1007/978-3-030-05971-2_2

2.2 Vibration Signal Examples

Vast majority of data on the condition of a wind turbine drivetrain is obtained via vibration signals. Let us present few examples of such signals. These examples will be used throughout the book to better visualize the introduced concepts.

Each digitized signal has two key parameters, namely the sampling frequency f_s and the number of samples N. The reciprocal of the sampling frequency is the time spacing between samples.

$$T_s = \frac{1}{f_s} \tag{2.1}$$

The sampling frequency in real applications varies from several kHz to over 100 kHz, though the most popular is 25 kHz. Thus, samples are taken every 1/25000 Hz = 40 μs.

The duration of the signal is given by:

$$T = N \cdot T_s \tag{2.2}$$

Depending on actual implementation in a software system, the duration may equal to $N - 1$ samples rather than N. This is the case e.g. in Matlab. In practice the total duration of a signal varies from 0.5 s to several minutes for low frequency signals. The duration time on the one hand greatly increases the requirements towards the CPU power and memory. On the other hand, one can obtain much better frequency resolution df, which is defined by the formula:

$$df = \frac{1}{T} = \frac{1}{N \cdot T_s} = \frac{f_s}{N} \tag{2.3}$$

Let's start with the simple sine waveform. It is a single sine function of frequency 8 Hz and amplitude of 1.8. Thus, the signal is defined as:

$$x(t) = 1.8 \cdot \sin(2\pi \cdot 8 \cdot t) \tag{2.4}$$

The signal is presented in the Fig. 2.1. This example will be useful in presenting frequency analysis tools later in this chapter.

In real mechanical systems signals are often composed of several harmonic components, i.e. signals which are a multiple of a fundamental frequency. The next example signal is composed of three sine waves, and is given by Eq. (2.5).

$$\begin{aligned} x(t) = {} & 1.3 \cdot \sin(2\pi \cdot 8 \cdot t + 0) + 0.6 \cdot \sin\left(2\pi \cdot 16 \cdot t + \frac{\pi}{2}\right) \\ & + 0.5 \cdot \sin\left(2\pi \cdot 24 \cdot t - \frac{\pi}{4}\right). \end{aligned} \tag{2.5}$$

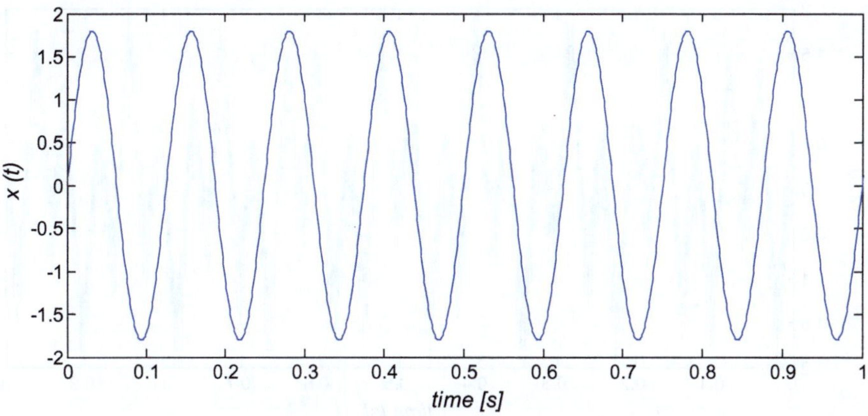

Fig. 2.1 Pure sine wave signal as given in Eq. (2.4)

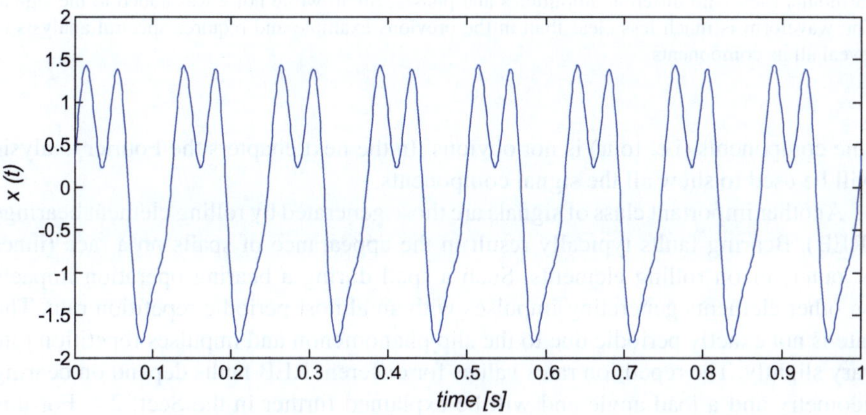

Fig. 2.2 Signal composed of three harmonics of a 8 Hz sine with different amplitudes and phases, as in the Eq. (2.5)

Please note that now the signal has the fundamental frequency (of amplitude 1.3 at the frequency 8 Hz), the second harmonic (i.e. 16 Hz of amplitude 0.6) and the third harmonic (i.e. 24 Hz of amplitude 0.5). Please, note different phase shifts of all the components (respectively, 0, $+\frac{\pi}{2}$ and $-\frac{\pi}{2}$ for the three first harmonics). The plot of the signal is presented in Fig. 2.2.

Another more complex example is created by simulating a signal from two shafts and a gearbox. The fast shaft has rotational speed of 25 Hz (1500 rpm), the slow shaft—of 9 Hz (540 rpm). Additionally, a component from a gear meshing was added, of 225 Hz together with its second harmonic of 450 Hz. There is also some white noise added. The plot of the signal is presented in the Fig. 2.3. The visual analysis of the signal is harder than in the previous example. Even the number of

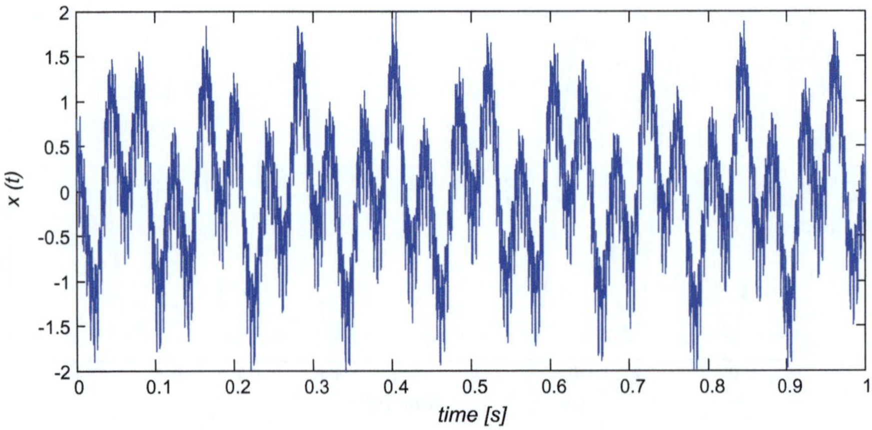

Fig. 2.3 Signal composed of sine signals from two shafts and gearmesh frequency with its second harmonic, each with different amplitudes and phases. Small white noise was added to the signal. The waveform is much less clear than in the previous example and requires spectral analysis to reveal all its components

sine components (i.e. four) is not obvious. In the next chapters the Fourier analysis will be used to show all the signal components.

Another important class of signals are those generated by rolling element bearings (REB). Bearing faults typically result in the appearance of spalls on a race (inner or outer) or on rolling elements. Such a spall during a bearing operation impacts on other elements generating impulses with an almost periodic repetition rate. The rate is not exactly periodic due to the slip phenomenon and impulses repetition rate vary slightly. The repetition rates values for different REB faults depend on bearing geometry and a load angle and will be explained further in the Sect. 2.4. For this example it is important to know that there is an additional component in a vibration signal, as shown in the Fig. 2.4. The signal was simulated as a series of decaying sine waves with a small white noise. From the signal processing point of view, the signal in the Fig. 2.4 is created by a carrier frequency (the sine wave) modulated by periodic impulses. The frequency of the sine wave (i.e. the carrier frequency) depends on a machine structure and it is much higher than the fault frequency (i.e. the modulating frequency). The sine wave used for illustration of a carrier frequency is a simplification as well, because in practice a high number of structural frequencies are excited by such repetitive impacts. It is important to note that in this type of signals the sought signal parameter is the repetition rate rather than the excited structural frequencies.

In practice the amplitude of this additional component, especially at an early stage of fault development, is very small and is not visible in the signal. That is why only the investigated component was presented in the Fig. 2.4 with a small white noise, but without any components from shafts, gearboxes etc. The signal containing both signals, i.e. with simulated components from shafts, gearbox and a faulty bearing, is presented in the Fig. 2.5.

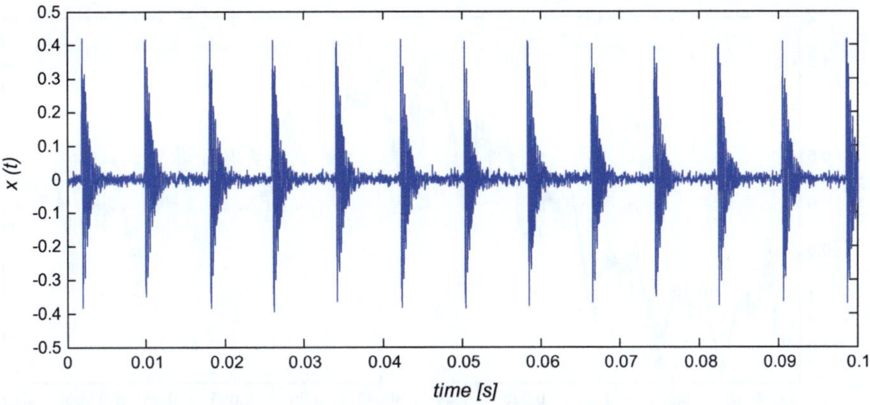

Fig. 2.4 Simulated signal showing a component generated by a faulty rolling element bearing. The sine carrier frequency is modulated by the periodic impulses. The fault signature are decaying impulses while the spacing between them is the fault repetition rate

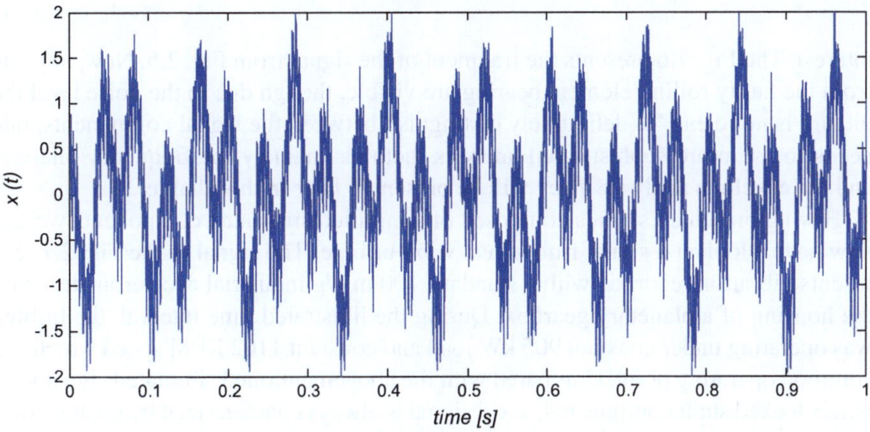

Fig. 2.5 Simulated signal which combines two shafts, a gearbox and a component from a faulty rolling element bearing. The impacts from the faulty bearing are too small to be noticed on a waveform and thus the figure is very similar to the Fig. 2.3

A signal waveform is simply its oscilloscope view, i.e. a plot of instantaneous value of the vibration signal in time. Example waveforms were presented in the previous chapter. The analysis of a waveform gives important insight into the signal. It is usually one of the first steps of its analysis.

The waveform analysis is particularly important when investigated signals are complex. It is the case when a signal contains short impulses, which is often the case in rolling element bearings faults. The example of such a signal was presented in the Fig. 2.5. Since different components in the signal vary in time and amplitude, it is important to be able to zoom the waveform plot in order to focus on the part of

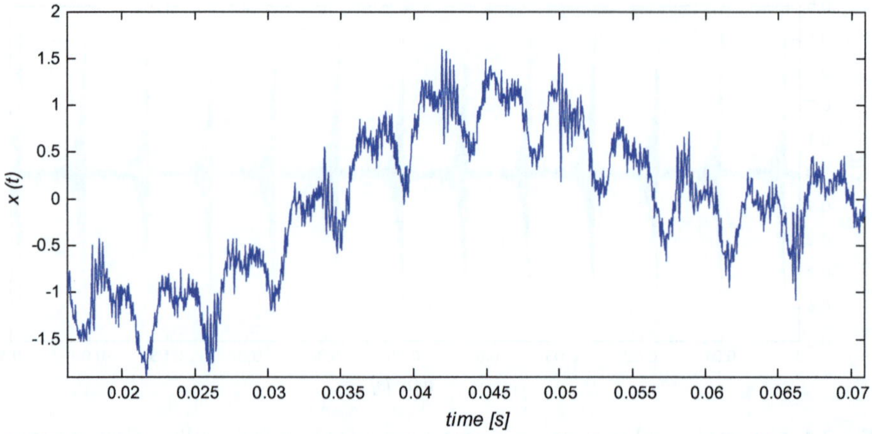

Fig. 2.6 Zoomed fragment of the Fig. 2.5. Now, the repetitive impacts from a faulty rolling element bearing are visible, though masked by noise

interest. The Fig. 2.6 presents the fragment of the signal from Fig. 2.5. Now, impacts from the faulty rolling element bearing are visible, though due to the noise level the picture is not clear. To definitively distinguish between the signal components, one needs to use more sophisticated analysis methods, namely the frequency analysis and the envelope analysis. They will be presented later in this chapter.

Having presented signals composed of simplified simulated components we can now have a look at a signal from a real wind turbine. The signal in the Fig. 2.7 represents vibration recorded with a standard 100 mV/g industrial accelerometer from the housing of a planetary gearbox. During the illustrated time interval, the turbine was operating under constant 905 kW load and constant 1102 RPM speed which is a common operating point. Compared with the abovementioned simulated signal with phase-locked shaft components, a real signal is always characterized by greater complexity of components and a lower signal-to-noise (SNR) ratio. The figure clearly illustrates a presence of a dominant low frequency component with a period approximately equal to slightly more than $3/5 * 0.01$ s $= 0.006$ s, i.e. $1/0.006$ s $= 166$ Hz.

Another example of waveform usefulness are signals from transient states, e.g. during start-ups. In such cases the structure of a signal can be very complex and the waveform analysis is the first step to isolate and analyze particular components.

The Fig. 2.8 presents a signal acquired during the stop of a wind turbine. The signal is very complex and consists of decaying vibration and several impulses. All the component frequencies decrease as well, making the signal analysis much harder. The exact nature of the signal or its frequency content cannot be visible. For such complex cases the waveform analysis alone is not sufficient and should be extended with the time-frequency analysis.

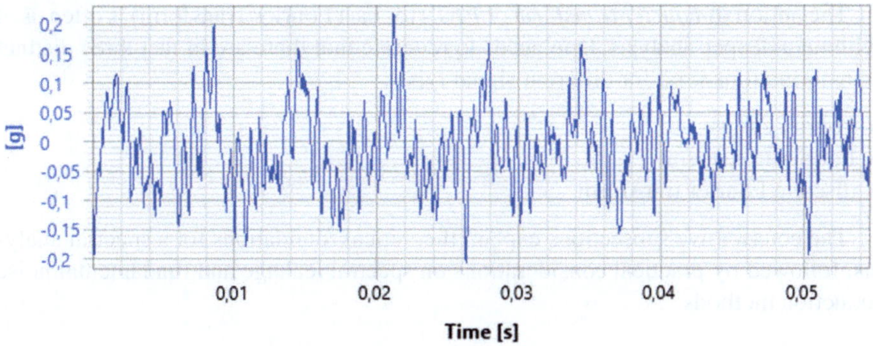

Fig. 2.7 Example of a real signal from a wind turbine planetary gearbox

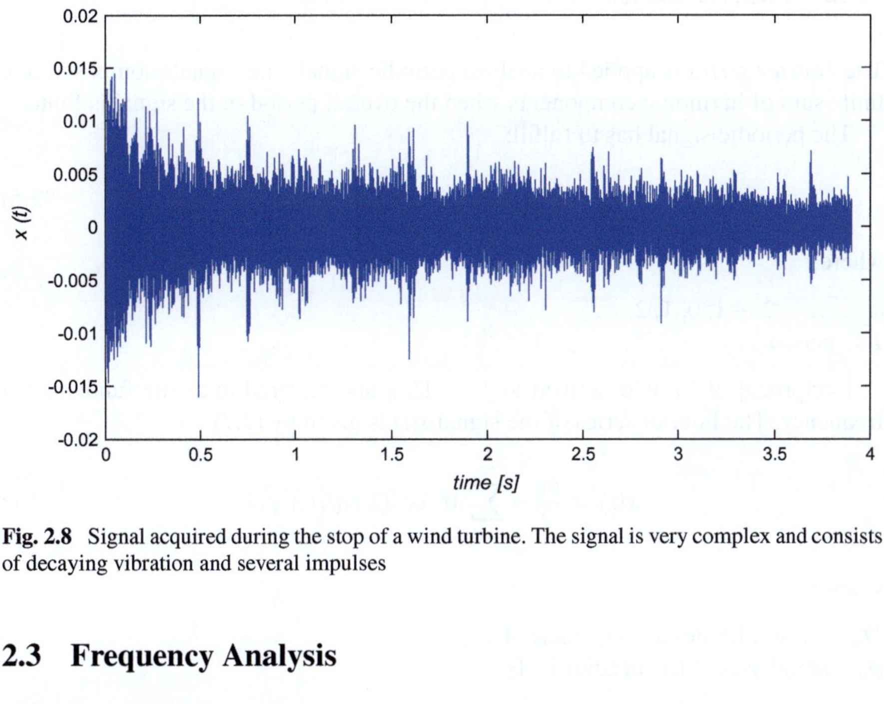

Fig. 2.8 Signal acquired during the stop of a wind turbine. The signal is very complex and consists of decaying vibration and several impulses

2.3 Frequency Analysis

2.3.1 Overview

The idea of the frequency spectrum is crucial to the rotating machinery analysis. With this approach, a frequency content of a signal is represented. Various spectral lines in such a representation can be further linked to particular drivetrain components, as each one has a distinctive rotating period and thus the source of an increased (or changed) vibration can be identified.

The notion of *Fourier transform* or *FFT* (for Fast Fourier Transform) is often used without a deeper analysis. One needs to reiterate that there are in fact three distinct transformations used for different signal types:

- Fourier series,
- Fourier transform,
- discrete Fourier transform.

The consecutive subsections explain theoretical foundations for a spectral analysis, followed by practical considerations on spectral leakage and fundamental noise reduction methods.

2.3.2 Fourier Series

The *Fourier series* is applied to analyze periodic signals, i.e. signals composed of a finite sum of harmonic components when the overall period of the signal is finite.

The periodic signal has to fulfill:

$$x(t) = x(t + nT_0) \tag{2.6}$$

where:

n ..., −2, −1, 0, 1, 2, ...
T_0 period

Reciprocal of T_0 is described as $f_1 = 1/T_0$ and referred to as the fundamental frequency. The Fourier series of the signal *x(t)* is given by (2.7):

$$x(t) = \frac{a_0}{2} + \sum_{n=1}^{\infty} M_n \cos(2\pi n f_1 t + \varphi_n) \tag{2.7}$$

where:

M_n are amplitudes at frequencies f_1, f_2, ...
φ_n are phases at frequencies f_1, f_2, ...

Another popular Fourier series representation is the *complex form*, given below (2.8), (2.9):

$$x(t) = \sum_{n=-\infty}^{\infty} c_n e^{j2\pi \frac{nt}{T_0}} \tag{2.8}$$

$$c_n = \frac{1}{T_0} \int_0^{T_0} x(t) e^{-j2\pi \frac{nt}{T_0}} dt \tag{2.9}$$

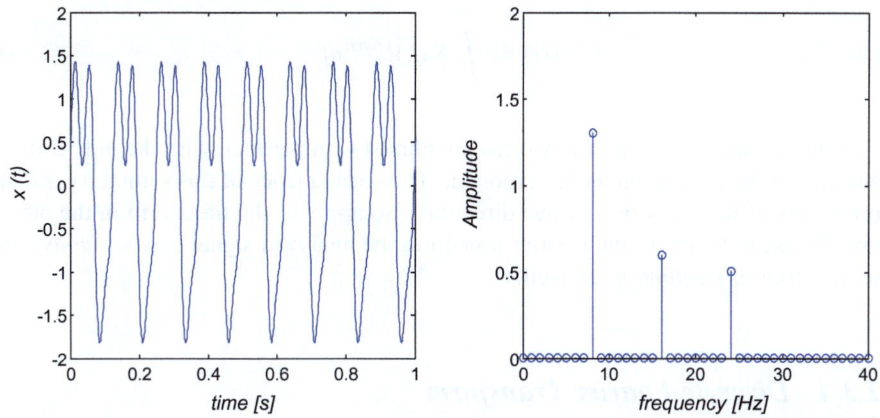

Fig. 2.9 Signal as given in Eq. (2.5) (left) and its Fourier series transform amplitudes (right)

Let's refer to the signal from Eq. (2.5) which is the sum of three sine functions of frequencies, respectively, 8, 16 and 24 Hz with different amplitudes and phases. The signal is continuous and with the Fourier transform it can be presented as a sum of harmonic components where the basic component is of frequency 8 Hz. Thus, the spectrum with such a transform can be understood as a plot of Fourier series complex coefficients amplitudes in the function of frequency. This plot only exists at multiples of 8 Hz. If a longer period is taken (a multiple of T_0), more coefficients are calculated, but their value is zero. In the Fig. 2.9 the period for the transform was 1 s which equals 8 periods of the signal from the Eq. (2.5). Thus, the first harmonic appears at the line number 8 rather than 1. The additional 7 periods padded the resulting Fourier series with zeros.

Please note that the analyzed signal is *continuous and periodic* and its spectrum is *discrete*. The example signal above is periodic only when ratios between signal frequencies are integer. If f_2 frequency is, e.g. $\sqrt{2} f_1$, than T_0 does not exist and the signal is so called almost periodic. Applying Fourier series will not yield correct results, because the assumption about periodicity is not met.

2.3.3 Fourier Transform

The *Fourier transform* does not assume that the signal must be periodic. It can be used to analyze signals which are periodic, almost periodic or random. The Fourier transform is given by:

$$X(f) = \int_{-\infty}^{\infty} x(t) \mathrm{e}^{-j2\pi ft} dt \tag{2.10}$$

$$x(t) = \int_{-\infty}^{\infty} X(f)e^{j2\pi ft}df \quad (2.11)$$

One can note that these transforms are almost symmetrical, with the only difference being the minus sign in the exponent. The consequence of this symmetry is that properties of the transform in one direction also apply to the transform in the other one. Please note that in the Fourier transform the analyzed signal is *continuous* and its spectrum is *continuous* as well.

2.3.4 Discrete Fourier Transform

Real signals in data acquisition systems are sampled, i.e. discrete, as we only know the signal value at sampling moments. Sampled signals are also limited in time, as we are not able to store and process infinite amount of data. It holds true, despite amazing growth of capabilities in data storage and processing. To obtain spectra of sampled signals, we use the *Discrete Fourier Transform (DFT)*, given by:

$$X(k) = \sum_{n=0}^{N-1} x(n)e^{-j\frac{2\pi}{N}nk} \quad (2.12)$$

$$x(n) = \sum_{k=0}^{N-1} X(k)e^{j\frac{2\pi}{N}nk} \quad (2.13)$$

where N is the number of samples in the sequence.

Again, please note the symmetry between the two transforms. The DFT is the most widely used analysis tool for vibration signals and forms the basis for many more advanced tools. The practical implementation of DFT is Fast Fourier Transform (FFT).

Let's consider the signal from the Fig. 2.3, simulating two shafts at 9 and 25 Hz with gearmesh at 225 Hz and its second harmonic at 450 Hz. It was not possible to analyze all these components on the waveform plot. The signal is sampled with the frequency of 100 Hz. Its spectrum reveals a periodic component within a noise. Frequencies and amplitudes are clearly visible (Fig. 2.10).

Please note that in DFT the analyzed signal is *discrete and periodic* and its spectrum is *discrete and periodic* as well.

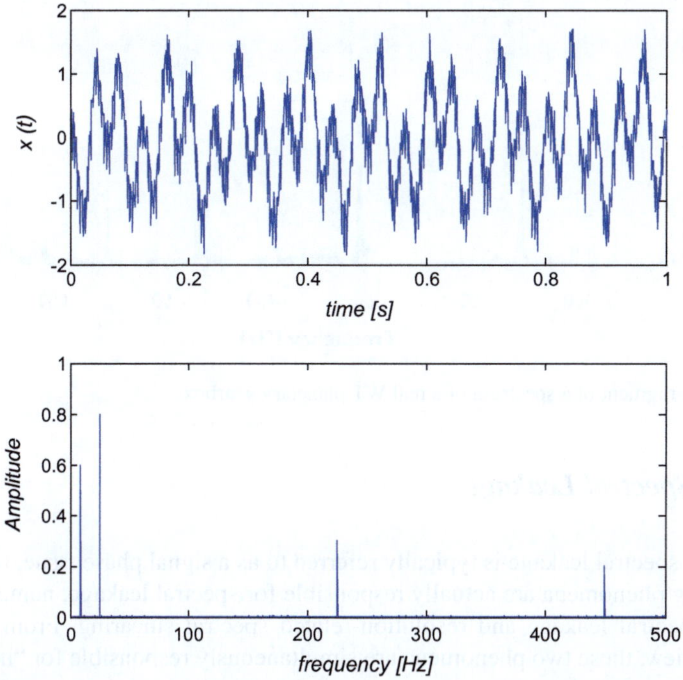

Fig. 2.10 Signal as given in example in the Fig. 2.3 (top) and its Discrete Fourier Transform amplitudes (bottom). All the signal components at frequencies 9, 25, 225 and 450 Hz are clearly visible

2.3.5 Real Signal Spectrum

A corresponding spectrum of the time signal in the Fig. 2.7 is illustrated in the Fig. 2.11. The figure shows dominant gear meshing components generated by a planetary gearbox.

As observed in the figure, the gear meshing frequency is accompanied by subsequent second and third harmonic components, and up to ca. 550 Hz range, these components dominate the spectrum. Although the GMF harmonics are clearly identified in a peak-sense, they do not constitute a major portion of a signal in energy (or power, RMS) sense. For this reason, condition monitoring of WT drivetrains is strongly based on tracking selected frequency (or order) bands related to particular drivetrain mechanical elements.

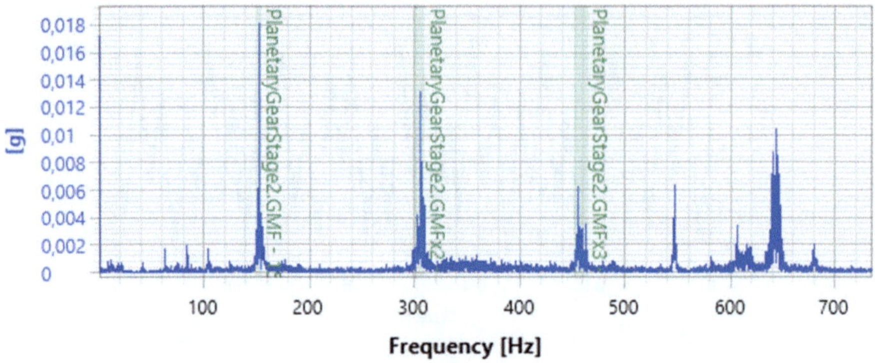

Fig. 2.11 Fragment of a spectrum of a real WT planetary gearbox

2.3.6 *Spectral Leakage*

Although spectral leakage is typically referred to as a signal phase issue, two signal processing phenomena are actually responsible for spectral leakage, namely phase-related spectral leakage and resolution-related spectral smearing. From practical point of view, these two phenomena are simultaneously responsible for "imperfect" spectral representation of frequency components.

The first phenomenon, typically called a "leakage" is a consequence of the fact that a particular frequency component has a different instantaneous phase at the beginning and at the end of the signal. The Fig. 2.12 illustrates a single sinusoidal component with frequency equal to 100 Hz. The sampling frequency is 1000 Hz, so in a 1 s signal, ten full and compete cycles are recorded.

While calculating the FFT, a signal is assumed to be repeated, as the transform assumes infinite duration of the signal. In the current example, during every repetition, the first point of the next instance perfectly matches the next point after the last point at the end of the signal. As a consequence, the signal is perfectly represented as a single frequency component. The Fig. 2.13, on the other hand, illustrates the Fig. 2.12 counterpart, but this time the component frequency is equal to 100.3 Hz, so 1 s of this signal does not contain integer numbers of cycles of a component with this frequency.

This time, time waveforms clearly show that the beginning and the end of the signal are not in phase, meaning there is a "jump" if the signal is to be repeated from either side (as assumed implicitly by the Fourier transform). The spectrum in the Fig. 2.13 shows that the 100.3 Hz component is represented by many components with the dominant component that is closest to the 100.3 Hz, i.e. 100 Hz. It is worth noting that for a non-smeared component the spectral maximum value is practically equal to the time peak, whereas for a smeared spectrum the dominant peak is nearly 15% smaller. The amplitude loss depends on a ratio between the frequency component in the signal and the exact position of spectrum lines. In an extreme case it can reach as low as 50% of a real signal amplitude.

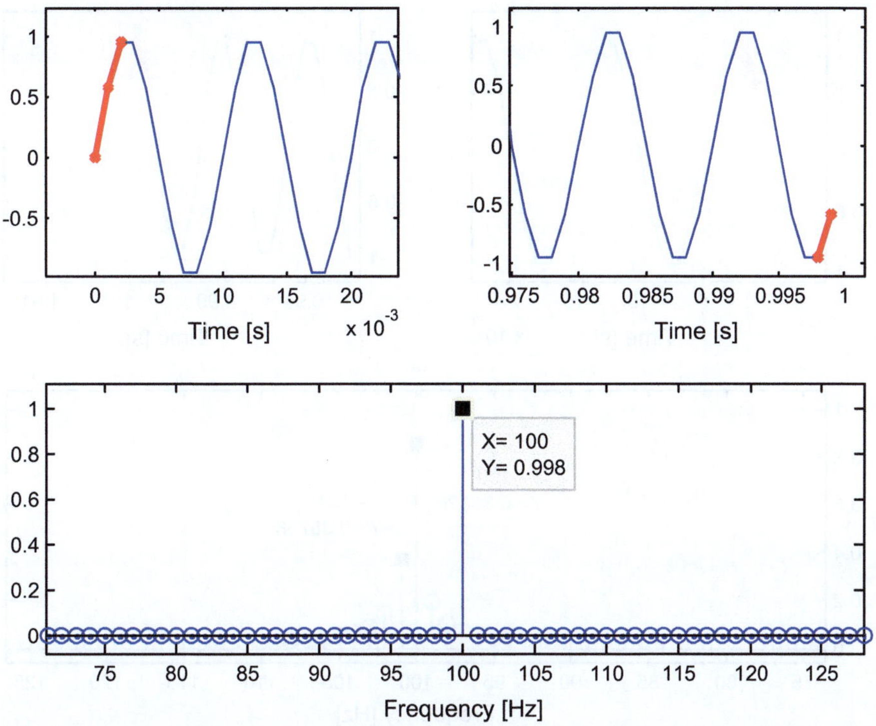

Fig. 2.12 Signal without leakage, top-left: beginning of the signal, top-right: end of the signal, bottom: spectrum of the signal

2.3.7 Signal Windowing

The most popular method of minimizing the effect of spectral leakage is to multiply a signal by such a windowing function which virtually brings the time waveform symmetrically to a low value at both the beginning and the end of a signal. Although many windowing functions have been presented, practically only a few are used. In case of vibration analysis in which windowing aims to limit spectral leakage, only these functions which bring a signal to zero are used. The Fig. 2.14 illustrates windowing functions of the following type:

- hann window,
- hanning window,
- hamming window,
- flattop window,
- rectangular window,
- tukey window (12%).

Particular windowing functions have been deliberately selected in order to point out a few practical remarks listed below:

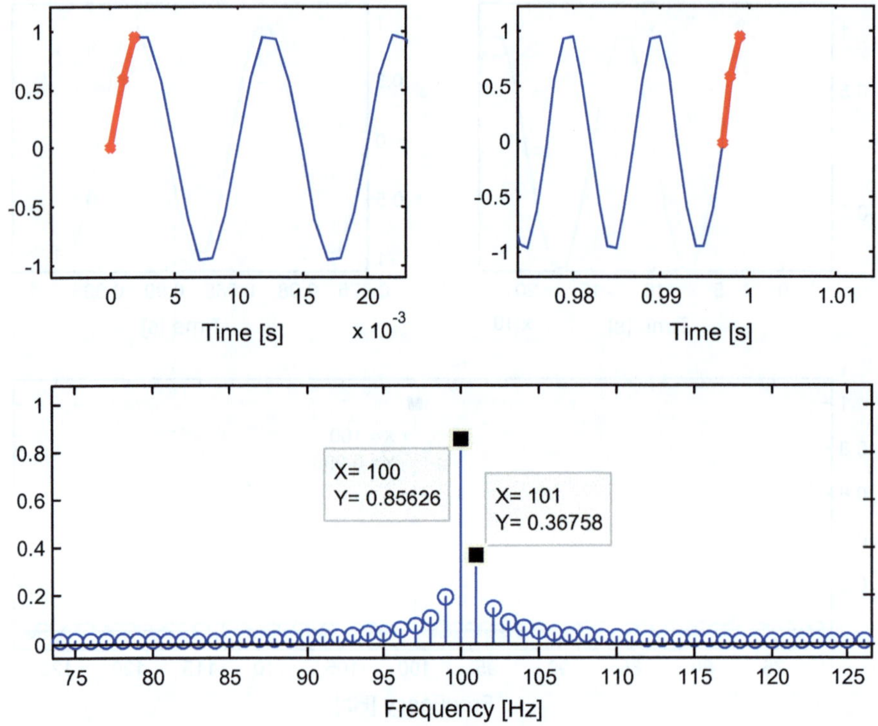

Fig. 2.13 Signal with leakage, top-left: beginning of the signal, top-right: end of the signal, bottom: spectrum of the signal

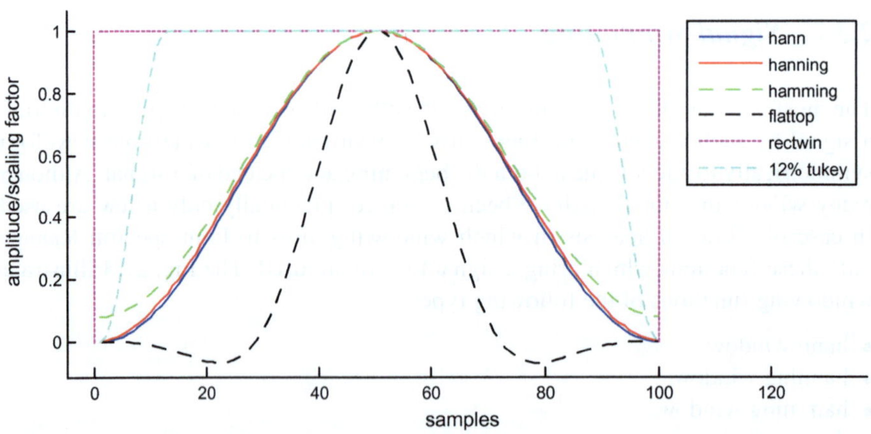

Fig. 2.14 Selected windowing functions

Hanning and Hann windows yield very similar results, but a true Hanning window has non-zero elements at boundaries, as illustrated in Table 2.1:

In many systems these two windowing functions are frequently treated interchangeably. Although from a practical condition monitoring point-of-view they are the same, according to the mathematical definition only the "hann" windowing function guarantees a full reduction of a phase "jump". Although the hann window brings a signal to zero, it never completely reduces the spectral leakage due to its inherent frequency spectrum sidelobes, which is an important point.

- Hamming window has significant non-zero boundaries and it does not fully solve the leakage problem,
- windows crossing zero, like "flattop", modify signal in the undesired manner,
- rectangular window is like no-window; therefore it is sometimes called a "default" window,
- tukey window is a general window which is equal to other windows for particular parameters; It describes the ratio of a cosine function to the rectangular window.

The Fig. 2.15 shows how the implementation of a hann window modifies a single, 10 Hz sinusoidal wave of amplitude 1, sampled at 1000 Hz. As clearly seen in the figure, the windowing function not only modifies amplitudes, but the shape of the signal as well. For this sinusoidal wave the RMS of the windowed signal is calculated as follows:

$$RMS = \sqrt{1/N \sum \left(x_i^2 \cdot \text{hann}_i\right)} \tag{2.14}$$

Table 2.1 Values of exemplary samples for hann and hanning windows

Window type/Sample no.	1	2	3	4	5
Hanning	0.25	0.75	1.00	0.75	0.25
Hann	0.00	0.50	1.00	0.50	0.00

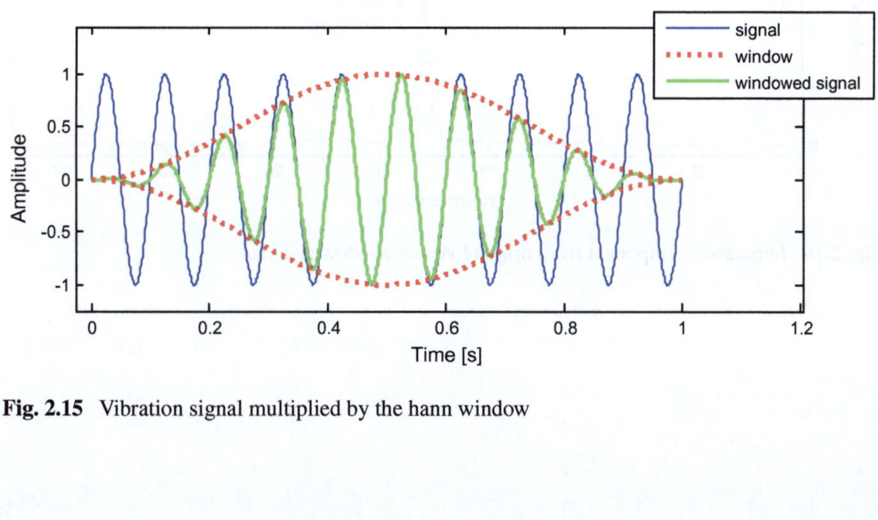

Fig. 2.15 Vibration signal multiplied by the hann window

which indicates that a windowed signal contains $0.4328/0.7068 * 100\% \approx 61\%$ RMS of the original one. For this reason, when diagnostic machine health indicators are compared against values taken from the norms, additional compensation coefficients should be implemented.

For instance, the RMS calculated from a modified signal from the Fig. 2.15 would be multiplied by a factor of 1.39 within the compensation process. Considering particular deterministic components in the frequency domain, implementation of windowing functions generally modifies the spectral peak in two ways, namely it increases the ratio of the peak value to the first sidelobe value (which is desirable) at the cost of total height of the peak value (which is not desirable). The Fig. 2.16 illustrates an imbalance component plotted form an original signal, i.e. a signal multiplied by a "rectangular" window against the same real signal modified by the hann windowing function.

As illustrated in the zoomed spectrum in the Fig. 2.16, the time signal is significantly reduced in its energy while the imbalance component is slightly narrower, yet significantly lower in amplitude, close to 50% of its original value.

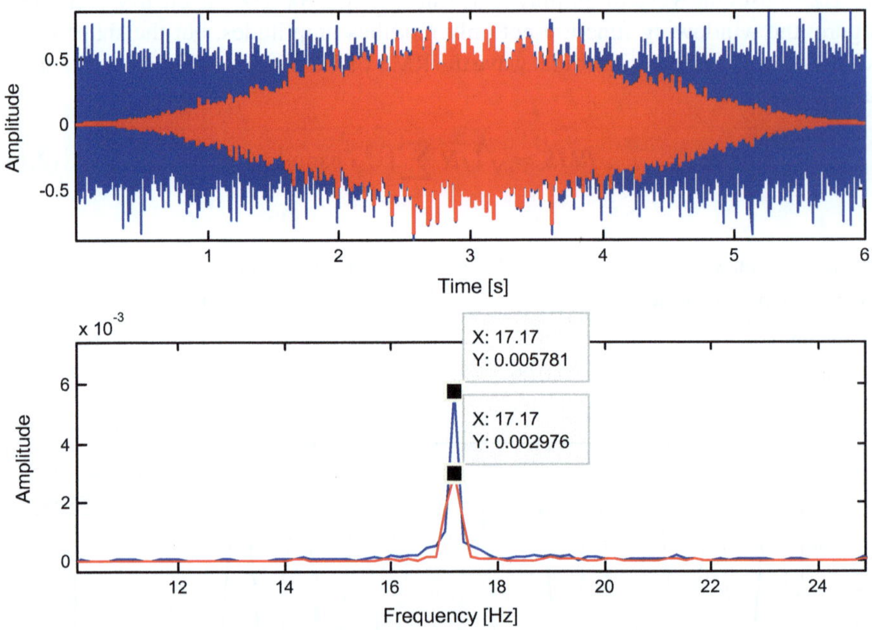

Fig. 2.16 Imbalance component from original versus windowed signal

2.3.8 Spectrum Digitization

The second phenomenon related to a spectral representation is the spectral resolution. As a reminder, a frequency spectrum might be interpreted as the measure of presence of some frequencies in a signal under investigation. The set of these frequencies is calculated as *0:df:fs/2*, where *df* is spectral resolution, and *fs* is sampling frequency. Spectral resolution is always equal to the reciprocal of the time of a particular record (i.e. signal length). So, for a 10 s signal, the resolution is $1/10 = 0.1$ Hz, for a 1 s signal, the resolution is $1/1 = 1$ Hz, for a 0.1 s signal, the resolution is $1/0.1 = 10$ Hz, etc. In this way, the set of "candidates" of frequencies to represent the frequency content of a signal is generated. Worth mentioning, this set does not depend on the sampling frequency which is a common mistake in the analytical deduction.

Considering the resolution as a reciprocal of the time T, we have: $df = 1/T$. Since time could be represented as sampling period multiplied by the number of samples, we have:

$$df = \frac{1}{T} = \frac{1}{dt \cdot N} \tag{2.15}$$

Sampling period is a reciprocal of the sampling frequency, therefore we can write:

$$df = \frac{1}{T} = \frac{1}{dt \cdot N} = \frac{fs}{N} \tag{2.16}$$

Finally, let us consider a scenario in which we are recording a signal with a fixed time T. For a given *fs*, we have N points. Now, if we increase the sampling rate by, for instance twice, we obtain a signal with twice as many points, therefore we do not modify the resolution:

$$df = \frac{1}{T} = \frac{1}{dt \cdot N} = \frac{fs}{N} = \frac{2 \cdot fs}{2 \cdot N} \tag{2.17}$$

According to a mathematical constraint of the Fourier analysis, this set of candidates is a set of orthogonal functions, which means that they are independent. Referring for example to the Fig. 2.13, this means that the candidate equal to 100 Hz is not related to the candidate 101 Hz, so they represent a portion of the 100.3 Hz component independently. From the resolution-point of view, if a particular signal component were to be represented completely by a single frequency component, a perfectly matching candidate, i.e. of the same frequency value would have to be present. However, in practical digital signal processing of WT drivetrain signals, time waveforms are recorded with popular lengths like 1, 10, 20, 60 or 100 s. As a consequence, the only case in which spectral smearing will not be present, is when WT generates signals with respective frequencies:

- 1 s: 1, 2, 3 Hz, …
- 10 s: 0.1, 0.2, 0.3 Hz, …

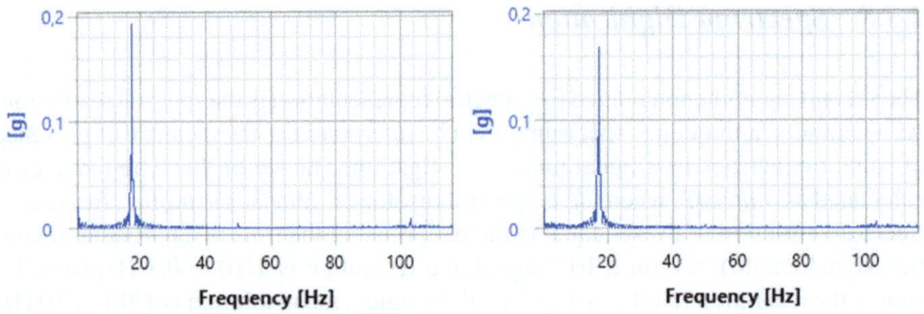

Fig. 2.17 Consequence of spectral leakage and smearing

- 20 s: 0.05, 0.1, 0.15 Hz, …
- 60 s: 0.016(7), 0.03(3), 0.05 Hz, …

For this set of candidates resulting from selected signal length, any frequency outside this set would be represented with smearing. The very first observation resulting from this fact is that when a full resolution spectrum is plotted real-time, amplitudes of both noise and deterministic components vary. The Fig. 2.17 illustrates spectra from 10 s vibration signals from the same machinery for constant speed and load recorded one after the other.

The first measurement shows an imbalance component with a peak value equal to ca. 0.185 g, whereas the other measurement shows the value of 0.165 g for the same fault. Practically, the peak (i.e. maximum) value of dominant frequency components might vary from measurement to measurement for a process assumed to be stationary by as many as 30%. In order to overcome this obstacle, vibration analysis offers windowing described in the previous chapter, spectral averaging, described in the next chapter as well as signal resampling, described in Sect. 2.5.

2.4 Envelope Analysis

2.4.1 Overview

The envelope analysis is a very powerful tool used primarily for early detection of rolling element bearing (REB) faults. This analysis is used when the signal of interest has a form of repetitive impulses buried in a much stronger signal. The Fig. 2.18 presents the layout of a rolling bearing under an unidirectional vertical load. The REB consists of: an inner race, an outer race, balls (or in general, rolling elements) and a cage, which holds rolling elements in a given relative position.

When a fault in one surface of a bearing strikes another surface, a force impulse is generated which excites resonances in a bearing and a machine. The examples of such signals were presented in the Figs. 2.4 and 2.5. Successive impacts produce a

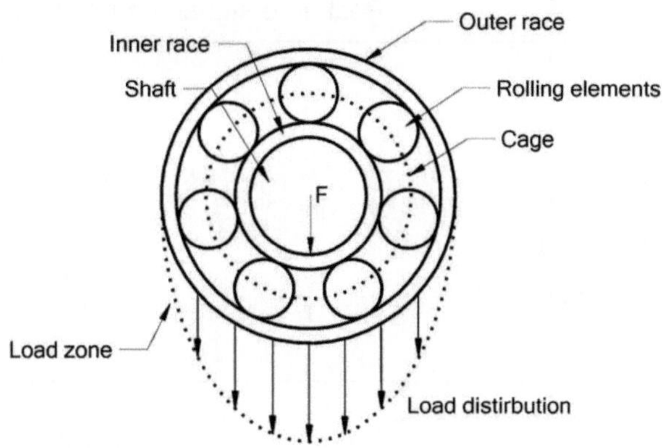

Fig. 2.18 Rolling element bearing components and load distribution. Rolling elements hitting a damaged component will generate periodic impacts

series of impulse responses which may be amplitude modulated as a result of the passage of the fault through the load zone or of the varying transmission path between the impact point and the vibration measurement point. The spectrum of such a signal would consist of a harmonic series of frequency components spaced at the bearing defect frequency with the highest amplitude around the resonance frequency. These frequency components are flanked by sidebands if there is an amplitude modulation.

Please note, that this model is only valid for a fault in its early stage of development, when it has a limited size. It is referred to as the *local fault*, as it is smaller than the rolling element and contact between the rolling element and the fault generates impulses at a known rate. In the later development of the fault its geometrical size grows and it becomes the *extended fault*. Then the size of the spall is roughly equal or larger than the rolling element and the excitation generated by the contact becomes much more complex. Individual impulses tend to mingle, but the overall energy of vibration increases. This is a measure which can help to distinguish between the local fault and the extended fault stage. When a fault reaches the extended stage the wear accelerates and a bearing should be replaced.

A theoretical vibration signal from a REB fault and its corresponding spectrum are shown in the Figs. 2.19 and 2.20. It is an idealized case which is valid only for a series of equally spaced force impulse excitations. In reality, there is a slip phenomenon, because REB components are not phase locked, i.e. there is a slight random fluctuation in the spacing between each force impulse, because the load angle on each rolling element changes as rolling elements enter and leave the load zone. This means the rolling diameter of each rolling element is slightly different and some will tend to roll faster than others. However, the cage keeps them apart at a certain mean spacing and they all travel around the bearing race at an average of the cage speed. Random fluctuation can be so small that the variation in the spacing may not be detectable by the naked eye. However, the effect on the spectrum is much more pronounced. Slight random fluctuation causes the frequency components to smear laterally, and at higher frequencies they may smear for more than one complete harmonic spacing.

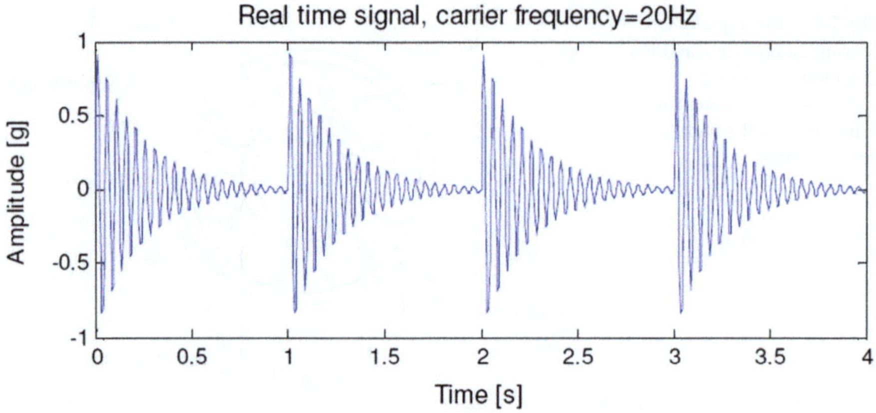

Fig. 2.19 Time series of pulse responses

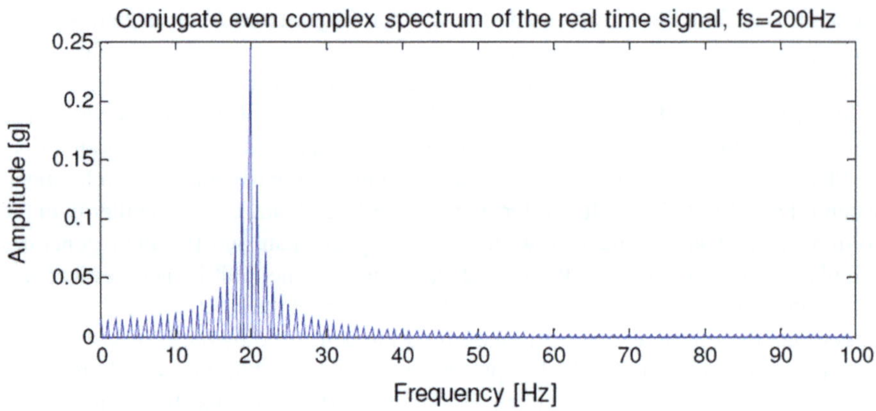

Fig. 2.20 Spectrum of the time series illustrated in Fig. 2.19

If one of the races experiences spalling, it will almost periodically impact with rolling elements. The fault signature is represented by successive impulses with a repetition rate depending on the faulty component, geometric dimensions and the rotational speed. The period between impacts is different for all listed elements and depends on the geometry of a bearing, the rotational speed and the load angle. The characteristic frequencies for a rolling element bearing (i.e. defect frequencies) were given in Chap. 1.

Impulses generated by a fault are usually described as pseudo cyclostationary to reflect the fact that the spacing between these impulses is not constant and usually varies between 1–2% as a result of slippage which—in turn—results from the variation of the load angle of a rolling element. The immediate consequence of this 1–2% variation (jitter) is smearing of the spacing between the repetition frequencies in the high frequency region. This makes detection of the fault extremely difficult using the spectrum of the raw signal, especially that the generated impulses are usually rather weak and are masked by other signal components, like gearmesh, low shaft harmonics, noise etc. The usage of the high frequency resonance technique (HFRT) by Darlow et al. [7], which is also widely known as the envelope analysis, and the usage of cyclostationary analysis presented by Randall et al. [8] and then by Antoni [9] and extended by Antoni [10], provide the base for fault detection and diagnosis in rolling element bearings. As it has been shown by many authors, e.g. Klein [4, 11], the envelope spectrum is a very efficient diagnostic tool for aforementioned faults, as the information about the fault is extracted from the spacing between impulses and not from the excited frequencies. The process of obtaining the envelope spectrum is often referred to as the signal demodulation. An informative source for rolling element bearing diagnostics can be recalled in a tutorial by Randall and Antoni [12].

The Fig. 2.19 illustrates a theoretical pulse response at 1 Hz repetition rate which could represent any of the characteristic frequencies listed in the table. The signal is theoretical, because it would be generated for an ideal point-mass systems, where the induced resonant frequency is a pure sine wave. In such a simplified case the resonant, or "carrier" frequency could be calculated from a time signal as reciprocal of the time between decaying HF sinusoidal peaks, as it is illustrated in Fig. 2.20.

The Fig. 2.20 illustrates that theoretically, every resonant frequency is amplitude-modulated by characteristic frequencies in the double-sideband (DSB) manner. In this case, the 20 Hz carrier frequency is clearly amplitude modulated (AM) by 1 Hz sidebands. Worth mentioning, these modulations are theoretically extended to both left and right boundary of the spectrum, but practically are quickly masked by noise.

Let's now analyze the signal from the Figs. 2.5 and 2.6. There is an additional component in a vibration signal, as shown in the Fig. 2.4. The signal was simulated as a series of decaying sine waves with a small white noise. The *envelope analysis* is applied to extract the information about the impulse repetition rate from the complete vibration signal, as shown in Fig. 2.4. The goal of the envelope analysis is the *demodulation* of the signal, which yields the modulating frequency, i.e. the characteristic frequency of the bearing fault. The Fig. 2.21 presents schematically key signal processing steps, showing both signals (left) and their spectra (right). On the plot (a), we can observe the original signal and its spectrum. The sought impulses are barely visible but the impulse resonant frequency can be seen in the frequency band of 5–6 kHz. On the plot (b), the signal was high-pass filtered to reject vibration signals from shafts and gears, which usually have much lower frequency. Now, the sought impulses become visible, but its spectrum has most of the energy in high frequency bands (as lower frequencies were discarded by the HP filter). At this step the actual demodulation is applied. The first step is signal rectification (thus the modulus symbol between rows (b) and (c). After this step new spectral lines appear in a low frequency band. It is so, because modulus is non-linear operation. The second step is applying a low pass filter which completes the demodulation, i.e. discards the high frequency part. After low pass filtering of the signal from Fig. 2.21c, we obtain the *envelope* of the signal—presented in the Fig. 2.21d. The information about

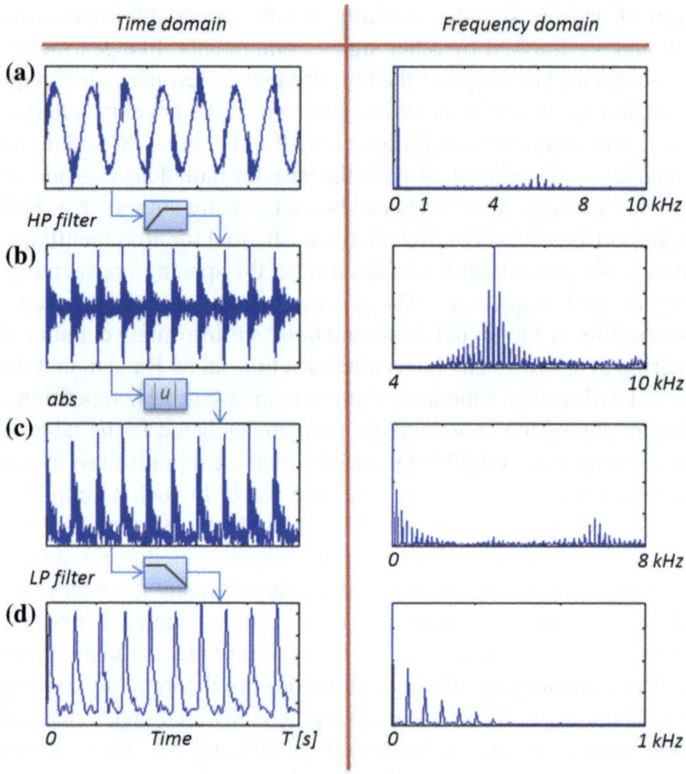

Fig. 2.21 Signal processing steps of the envelope analysis. Waveforms of signals are shown on the left, and their respective spectra—on the right

the carrier frequency is filtered out, and the shape (i.e. envelope) of the impulses is preserved. On the frequency spectrum one can finally see the spectral line at the fault characteristic frequency and its several first harmonics.

Another important feature of REBs induced impacts is the following: if a bearing load is constant and the outer race is stationary, the BPFO impulses will be similar to each other, because conditions for passage of each rolling element are uniform. This is not the case for the inner race fault and the rolling elements fault as they experience a varying load when moving in and out of the load zone, thus causing modulation at the shaft speed for the inner race case and at the cage speed for the rolling element fault case. An inner race case fault signature is presented in the Fig. 2.22 for both the time domain (top) and the envelope spectrum (bottom). In the Fig. 2.22 at the bottom, the harmonics of the BPFI as well as sidebands at the shaft speed are observed. A similar pattern will be generated in the case of the rolling element fault, but in that case the impacts would be modulated by the cage speed (or fundamental train frequency—FTF).

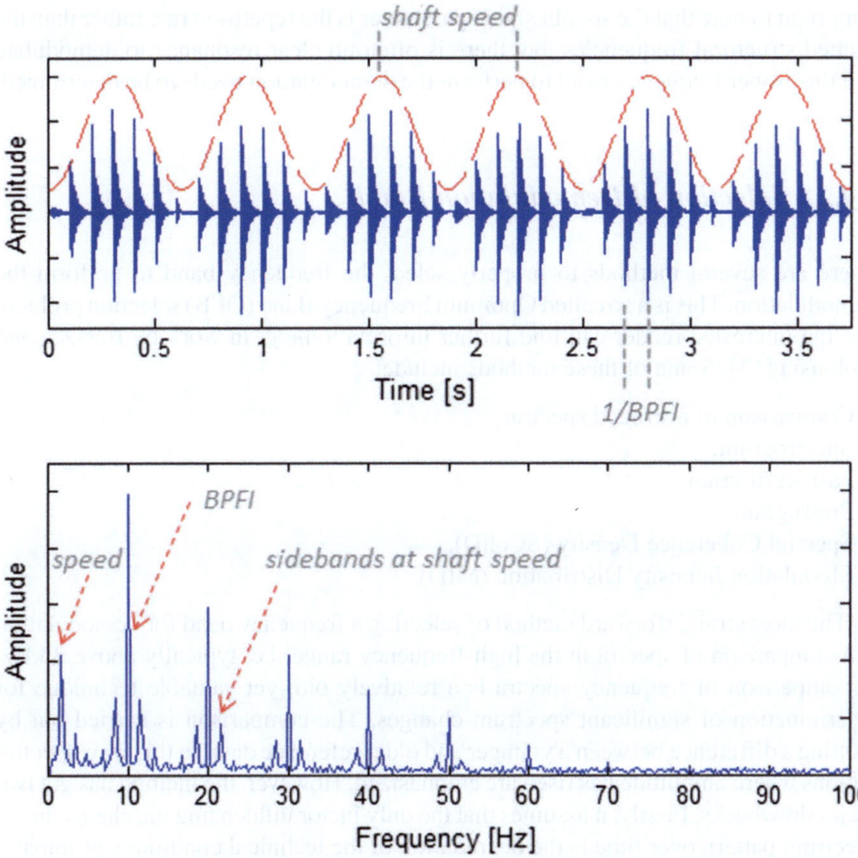

Fig. 2.22 Inner race fault: (top) time domain signature, (bottom) envelope spectrum

The most important question in the envelope analysis is the frequency band, in which the signal should be demodulated. In engineering practice, typically it is sufficient to only filter the signal with a high pass filter. The cut-off frequency should be above the second—third gear mesh harmonic of the fastest gear. It is the gear between the fast shaft and the intermediate shaft and the cut off frequency is in the range of 2 kHz. When the fault is in its early stage of the development or there are some other sources of impulses masking the sought signal of interest, the demodulation band should be narrowed. Such a method is called Narrowband Envelope Analysis (NEA). The best results are achieved when the band selected for demodulation is the band excited by the impacts in the faulty bearing, including the transfer path from the place of impacts to the sensor. In general, the frequency band of the induced impulses depends on a machine structure (and is much higher than the fault frequency). The sine wave used for illustration in the Fig. 2.20 is a simplification. In reality, a high number of structural frequencies are excited by such repetitive impacts. It is again important to note that the sought signal parameter is the repetition rate rather than the excited structural frequencies, but there is often no clear resonance to demodulate and the proper frequency band to perform the demodulation needs to be determined.

2.4.2 Selection of Demodulation Band

There are several methods to properly select the frequency band to perform the demodulation. This is a so called Optimum Frequency Band (OFB) selection problem and the interested reader can find further information e.g. in work by Barszcz and Jablonski [13]. Some of these methods include:

- Comparison of averaged spectra,
- Spectrogram,
- Fast Kurtogram,
- Protrugram,
- Spectral Coherence Density (SCohD),
- Modulation Intensity Distribution (MID).

The most straightforward method of selecting a frequency band for demodulation is a comparison of spectra in the high-frequency range, i.e. typically above 2 kHz. A comparison of frequency spectra is a relatively old, yet valuable technique for determination of significant spectrum changes. The comparison is carried out by plotting a difference between a younger and older reference data. In this way, spectral regions where amplitude has risen are emphasized. However, the method has got two major drawbacks. Firstly, it assumes that the only factor influencing the change in the spectrum pattern over time is the degradation of the technical conditions of impact-inducing machine elements. This is not true, for instance, if a machine part was replaced with another one characterized by different dynamics, therefore changing the overall frequency response function of a system without fault relevance.

Secondly, the resultant "difference" depends on the technique used for a spectral representation of the same signal from rotating machinery. If a full-resolution FFT is used, the resultant difference is inevitably biased with theoretically a 100% random error because the signal generated by impacts is random in nature. Therefore, the FFT comparison yields in practice significant results only for clear, significant spectral changes.

In order to overcome this problem, various averaged versions of spectral representation are used, most typically PSD (Power Spectrum Density), CPB (constant percentage spectrum) or other forms of a so-called "averaged periodogram". When using PSD, the signal power is practically averaged over frequency bins with resolution greater than the full frequency resolution ($1/T$ or fs/N). Depending on a particular system, these bins are either called "user spectral resolution" (in Hz) or "number of spectral lines". Due to the averaging process, a PSD is characterized by a greater signal-to-noise ratio (SNR) at the price of the spectrum resolution.

A spectrogram is one of the most common time frequency analysis maps which illustrates how the spectral density of the signal varies over time. The technique might enable to select the frequency range in which the damped impulses from local faults occur. A spectrogram might be calculated from a sequence of band-pass filters or via STFT (short-time Fourier transform). The latter is accomplished by breaking up the time signal into intervals (usually overlapping), and calculating the power spectrum for each part. The main drawbacks of the method include: (i) the knowledge of the number of intervals into which the signal is to be divided, (ii) complicated and time-consuming calculations, (iii) frequently challenging interpretation of the resultant color map due to relatively low power of the sought components in comparison with major signal components.

An optimal frequency band for signal demodulation might be selected by so-called Kurtogram as well. Fast Kurtogram was proposed by Antoni (2007). The tool illustrates kurtosis-based estimator values representing a peakiness of the filtered envelope time signals for a defined combination of center frequencies and band-widths. The result of the tool for a single vibration sample is a 2D colormap, where colors represent the kurtosis levels. The main drawback of the method is a high sensitivity to random extraneous components which can give ambiguous results.

Another version of a kurtogram-based method was presented by Bartelmus and Zimroz [14], where the optimal center frequency (indirect) detection algorithm is divided into "starting" and "ending" frequency. Even though the presented results are satisfactory, once again the kurtosis-based estimator shows a tendency to point out misleading frequency intervals, as stated by the authors. One of the solutions is the synchronous averaging, but it may cause the loss of information, too. A detailed description of exemplary use of spectral kurtosis within the process of diagnostics of a wind turbine drivetrain will be presented in the Chap. 5.

An optimal frequency band could be quite precisely selected with the use of so called "Protrugram" introduced by Barszcz and Jablonski [15]. The method displays the value of kurtosis of amplitudes of positive frequencies of a number of narrow-band envelope spectra. This solution has the disadvantage of gathering all the signal signatures "into a single pot". If the envelope spectrum is not masked by super-fluous components, e.g. fundamental harmonics, the method works fine. However, in many cases, the envelope spectrum (even narrowband) will include superfluous components which may ruin the Protrugram algorithm. A detailed description of this method will be presented in the Chap. 5.

2.4.3 *Considerations of Envelope Implementations*

Within practical implementation in condition monitoring systems, the calculation of signal envelope is characterized by two aspects, namely the demodulation technique and the demodulation frequency range. The demodulation techniques typically used include:

- classical path (HP filter, rectification, and LP filter),
- implementation of Hilbert transform in the frequency domain.

The first method is attractive for analog implementation, whereas the second one is characterized by perfect cut-off filter frequencies.

On the other hand, demodulation frequency range is typically selected in two ways, namely:

- fixed predefined range from 2–4 kHz up to system's Nyquist frequency,
- custom narrowband range.

The fixed predefined range is simple to configure and has clear interpretation. On the other hand, it often bears lower efficiency and is less sensitive to faults in its initial damage phase. The results may have false alarms if the vibration signal is contaminated by frequency inverter signatures. This first method is far more popular one in commercial systems. The main advantage of this method is its simplicity in configuration and traceability. The main disadvantage of the method is a relatively low efficiency comparing to the envelope signal calculated from an optimally selected frequency band. The Fig. 2.23 illustrates the envelope spectrum calculated using a standard fixed range (1000–12,500 Hz) and the Fig. 2.24—the envelope spectrum calculated with a frequency band optimized by the Protrugram tool (which will be described in the Chap. 5). The custom demodulation band has exceptionally high efficiency, but requires supervised expert adaptation and thus it is unsuitable for an automated monitoring process.

The Figs. 2.23 and 2.24 illustrate two envelope spectra from vibration signals collected with a standard industrial 100 mV/g accelerometer located in radial direction on a housing of a rolling-element bearing. The bearing experienced a local outer race fault with characteristic frequency equal to 5.79 order. The Fig. 2.23 clearly illustrates

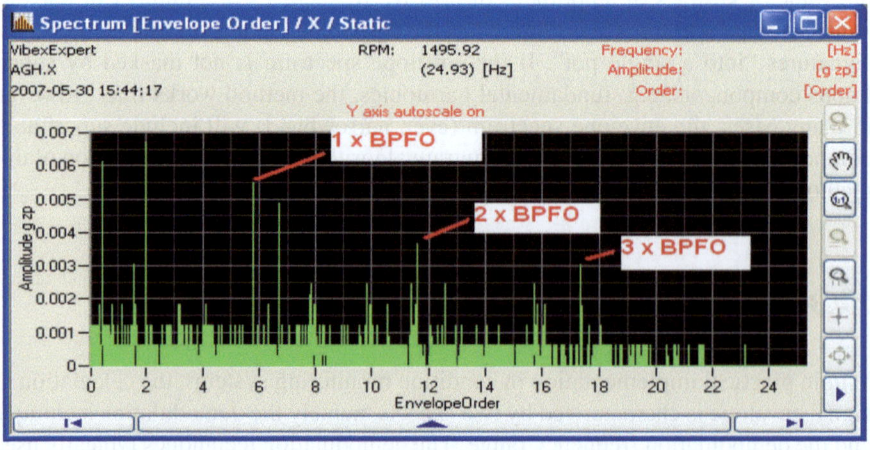

Fig. 2.23 Envelope spectrum calculated with standard commercially used parameters (1000–12,500 Hz)

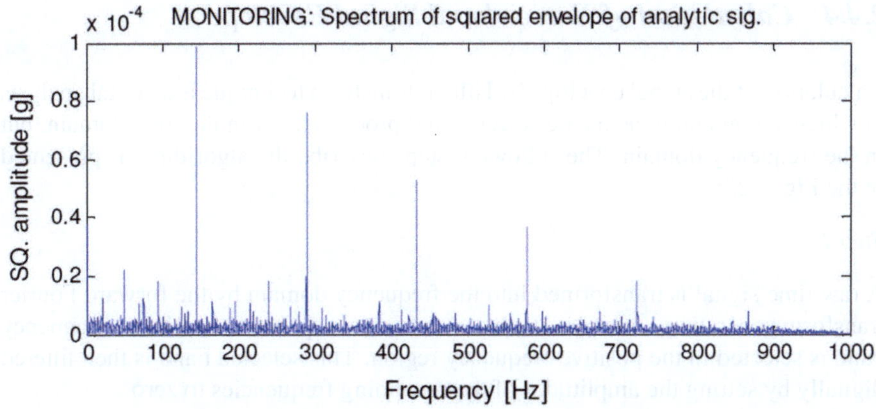

Fig. 2.24 Envelope spectrum calculated with expert-selected demodulation band supported by the optimization tool. The base signal was the same as in the Fig. 2.23

the first 3 harmonics of this frequency. The envelope spectra have been calculated by demodulating the signal initially filtered with a HP filter with a fixed cut-off frequency equal to 1000 Hz (in this case, up to 12 kHz). The consecutive harmonics of this frequency are not visible because this commercial system has implemented fixed LP cut-off frequency equal to 500 Hz, which is 500/current speed = 500/24.93 Hz = 20.05 order, so the 4th and consecutive harmonics are filtered out.

On the other hand, the Fig. 2.24 shows the envelope spectrum calculated via Hilbert transform in the frequency domain, demodulating the signal in the frequency band from approximately 3.5–4.5 kHz, as proposed by the Protrugram optimization tool. The characteristic frequency is equal to 5.79 order times 24.93 Hz the current rotational speed, which in this case gives $5.79 * 24.93 = 144.4$ Hz. In this case, the fundamental BPFO plus the consecutive 5 harmonics are clearly visible indicating a nearly book-case fault spectral representation.

Apart from resolution, the main difference between both graphs is absence and presence of the envelope components other than the BPFO harmonics, respectively. Additional components in the envelope spectrum calculated by the commercial system are naturally a consequence of demodulation of a much larger frequency band which corresponds to the bandwidth of the HP filter, independently of the subsequent LP filter. From mechanical point of view, any evident frequency apart from BPFO harmonic visible in the upper graph (e.g. 2.0 order) indicates a repetition rate of pulses which induce a resonant frequency in any range from 1 to 12 kHz. Taking advantage of the upper figure, this interval could be further limited to two ranges, namely 1–3.5 kHz and 4.5–12 kHz because if the induced resonant frequencies were in the 3.5–4.5 range, the 2.0 order (actually they are $2.0 * 24.93 = 49.86$ Hz) would be present in the upper figure as well, but they are not.

2.4.4 Calculation of Narrowband Signal Envelope

Calculation of the signal envelope by Hilbert transform technique is a digital analysis in which signal components are selected and processed not in the time domain, but in the frequency domain. The following steps describe the algorithm, as presented in the Fig. 2.25:

Step 1

A raw time signal is transformed into the frequency domain by the forward Fourier transform generating a complex double-sided frequency spectrum. Next, a frequency band is selected in the positive-frequency region. This selected band is then filtered digitally by setting the amplitudes of the remaining frequencies to zero.

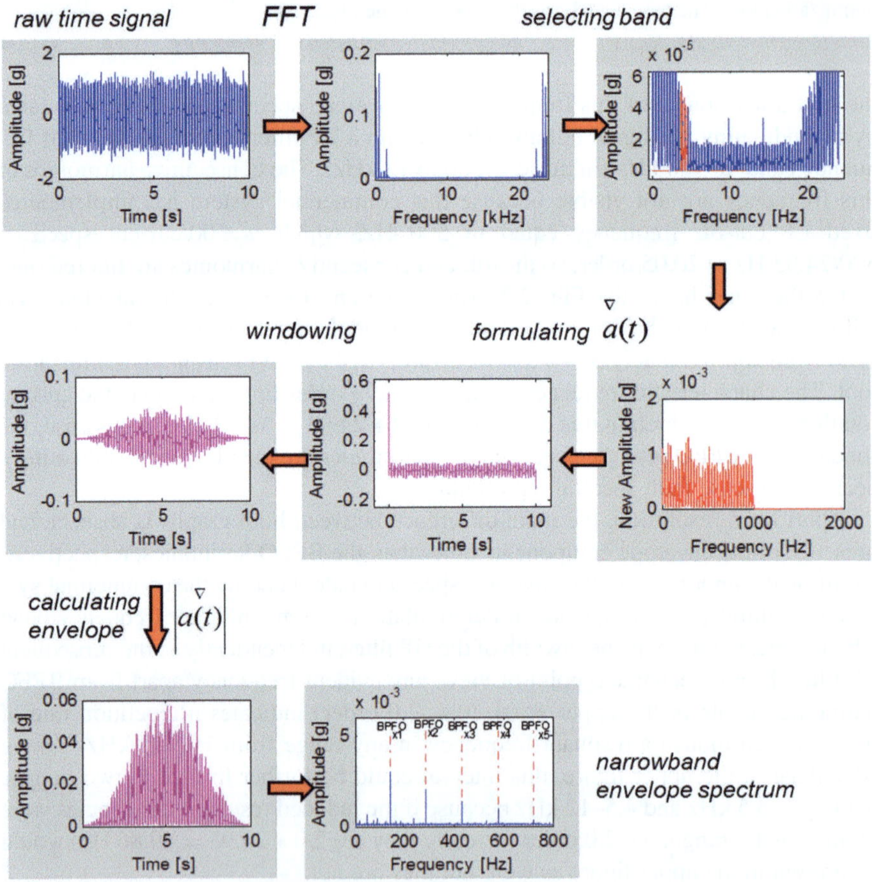

Fig. 2.25 Graphical illustration of the narrowband envelope analysis using Hilbert transform

Step 2

Next, the selected bandwidth is shifted "all the way to the left" meaning that the lowest frequency component of the selected bandwidth is at the low-frequency end (corresponding to the frequency of a DC component, f = 0 Hz). As a reminder from the signal processing theory, a phase shift of $-\pi/2$ at each positive frequency and $+\pi/2$ at each negative frequency introduced by Hilbert transform can be obtained by cancellation of negative frequency components and doubling the positive frequency components. It ensures a one-sided spectrum.

Step 3

Thus, the amplitudes of positive frequency components (i.e. of the selected bandwidth) are doubled, whereas the amplitudes of negative frequency components are set to zero. Next, the inverse Fourier transform is applied to the complex one-sided spectrum in order to obtain a complex analytic time signal. The analytic signal generated from a real vibration signal ought to be multiplied by a windowing function, preferably Hanning or Kaiser. The reason for windowing is that Hilbert transform of a signal composed of many components acts like a FIR filter, often called a "Hilbert transformer". Its impulse response extends infinitely in both direction. Since a digital infinite signal must be truncated "somewhere", the filter characteristics can be improved by applying a window function. Moreover, a practical Hilbert transform is causal (i.e. only for $t >= 0$). The interested reader can find exhaustive explanation in Stearns [16].

Step 4

The envelope time signal (real) is obtained from the complex one-sided spectrum by calculating the absolute value of the complex analytic time signal. Note that in order to calculate the absolute value of a complex analytic signal, a complex conjugate of that signal must be calculated. Depending on the software used, it can be done in three or in one step. Though it may seem trivial, it is of crucial importance to study consecutive steps. When a complex signal is multiplied by its conjugate, e.g.:

$$(a + jb)(a - jb) = a^2 - ajb + ajb - j^2b^2 = a^2 + b^2 = c^2 \tag{2.18}$$

The result represents square of the amplitude of the complex number. In order to obtain the amplitude itself, a square root must be taken. If a single complex number is of interest, this operation is straightforward and meaningless. However, in case of a real physical signal composed of many (co)sinusoids and their harmonics, square rooting changes the signal. In order to avoid the degradation (i.e. the modification) of a time signal, considering a squared envelope signal is preferable.

Step 5

Finally, the envelope spectrum of the real envelope signal of the complex analytic signal is obtained directly by a forward Fourier transform.

The presented algorithm illustrating the technique was shown as a very effective tool for diagnosis of a rolling element bearing. The concept of processing only a selected band of frequencies gives outstanding results in comparison with traditional enveloping techniques in which filters were applied to a time signal.

Nevertheless, the presented technique requires knowledge of the frequency band, where the signal generated by a faulty element is present. Generally, characteristic frequencies of bearings are said to be of range 5–20 kHz. Including such an interval for analysis makes the method rather low sensitive. As it was mentioned earlier, the quality of extraction of the modulating signal via Hilbert transform depends on the bandwidth of the carrier signal, which has been discussed in detail.

2.5 Order Analysis

Order analysis (also referred to as signal resampling which makes the important part of the method) is a method of analyzing the vibration signals generated by machines of varying rotational speed. All the fault detection and identification methods based on characteristic frequencies listed in the Chap. 1 assume that a faulty component generates a vibration signal at a specific frequency. These frequencies can be referred to a shaft fundamental frequency, like gearmesh frequencies, rolling bearing frequencies and then located on a frequency spectrum. Vibration levels of these spectral lines and their harmonics are related to the technical state of the particular component. Therefore, most diagnostic methods use comparison of vibration levels at characteristic frequencies with a baseline values or monitoring of trends of characteristic frequencies amplitudes.

The method presented above requires the frequency of the monitored spectral line to become constant. It is true as long as the rotational speed of the machine remains unchanged. As described in the Chap. 1, control of wind turbines is of two fundamental types: stall and pitch control. Stall controlled turbines operate on two distinct rotational speeds, 1000 and 1500 rpm (on the European market). This causes frequencies of the spectral lines to increase by 50% when the rotational speed is changed by the turbine controller, apart from short transient states. Pitch controlled turbines constantly change the pitch angle of the rotor blades depending on the wind speed and consequently the rotational speed of the turbine changes smoothly. The Fig. 2.26 presents the wind speed and the resulting rotational speed for a pitch controlled wind turbine. The speed varies between 860 and 1120 rpm, which is more than 25% of the average value (1000 rpm).

Such changes of the rotational speed have a large impact on the frequency spectrum. The Fourier transform over a given time period actually averages all the frequency components and yields a spectrum averaged on the analyzed time series. The Fig. 2.27 presents the frequency spectrum obtained from a vibration signal from a wind turbine with a constantly varying rotational speed. As a result of averaging, spectral lines are blurred and it is not possible to identify individual spectral lines.

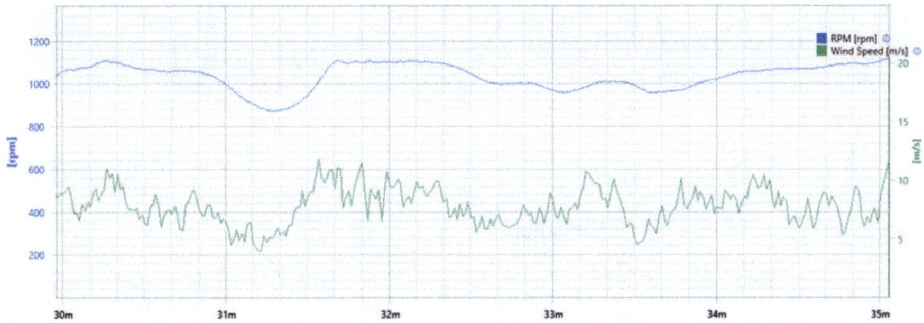

Fig. 2.26 Trend plot of wind speed (bottom line) and the resulting rotational speed (top line) of a pitch controlled wind turbine over a 5 min period. The range of rpm changes is close to 25%

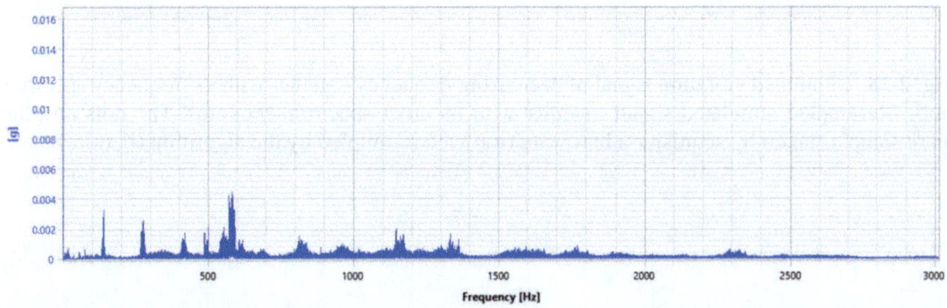

Fig. 2.27 Frequency spectrum for vibration data from a pitch controlled wind turbine. Spectral lines are blurred, especially for higher frequencies

Vibration signals taken from varying rotational speed machinery can be analyzed with a great accuracy, using the method of *order analysis*. The method requires that the time signal should be *resampled* before performing the frequency analysis.

The idea of the order analysis is presented in the Fig. 2.28. Most often, analogue signals are sampled with a constant sampling frequency, i.e. the time distance between consecutive samples is constant. For order analysis it is required that the number of samples is constant per one revolution of a shaft, and not per a time period. Then, despite changing signal frequency, the corresponding spectral line remains unchanged, as presented in the Fig. 2.28. Dots on both plots mark samples which are then used by the FFT algorithm. On the top plot the dots mark the samples taken by the data acquisition hardware and are spaced by the same period of time. On the bottom plot the dots mark the samples calculated by the resampling algorithm. They are not equidistant anymore, but there is a constant number of dots per one shaft revolution.

Since after resampling the samples are not any equidistant in time anymore, the domain after the Fourier transform is not frequency either. It is called the *order domain*, where 1 is the rotating frequency of the reference shaft. Note that the horizontal axis on the right bottom plot in the Fig. 2.28 is called order, not frequency. Order can be understood as a harmonic of the reference shaft. The advantage of the

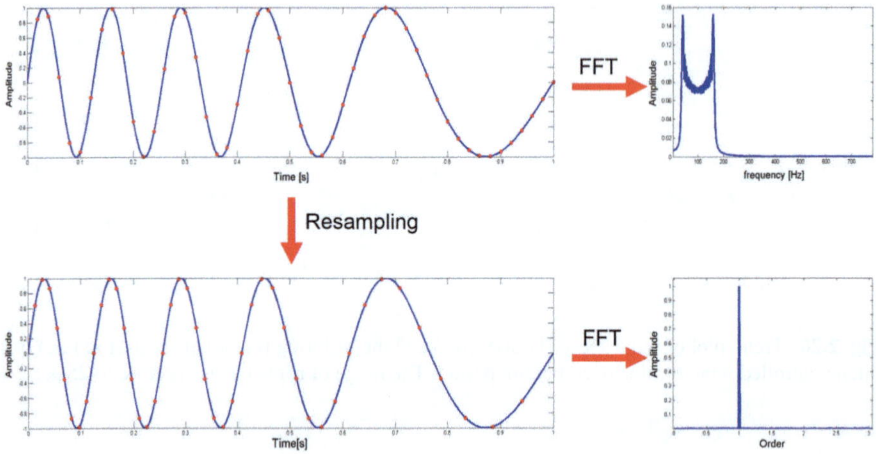

Fig. 2.28 Simulated vibration signal of decreasing frequency together with its frequency spectrum (top). Resampled simulated signal together with its order spectrum (bottom). The dots on plots mark samples taken by acquisition hardware (top) and resampled by the algorithm (bottom)

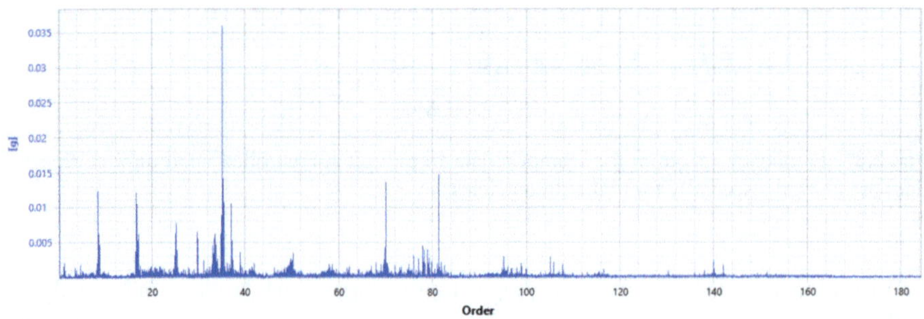

Fig. 2.29 Frequency spectrum for vibration data from a resampled signal from the Fig. 2.27. Spectral lines are now very sharp and it is easy to exactly identify the mechanical components related to a spectral line. Note that the peak amplitudes are much higher compared with the non-resampled signal

order domain is independence of order lines from rotational speed changes. Characteristic components of machinery faults are directly related to the machine speed and they stay constant in the order domain. The Fig. 2.29 presents the vibration signal after the resampling. One can observe that spectral lines are now much sharper, i.e. the signal energy is not dispersed on many spectral lines. It is easier to exactly identify the mechanical components related to a spectral line. Thanks to reduced energy dispersion the peak amplitudes are much higher compared with the non-resampled signal.

It is important to analyze how sampling with an equal number of samples per revolution can be achieved. It can be either performed in hardware, as it used to be done in the first designs. A hardware circuit which will take the rotational speed as the input and multiply it by a factor of several hundred is required. This method requires additional hardware, but reduces the requirements towards the CPU and memory. The modern method uses the phase marker signal (once per revolution) and the samples are calculated within the CPU. The resampling algorithm estimates the values of the vibration signal at time instants equidistant in the shaft angle domain. The interpolation used for resampling affects greatly the accuracy of results. The optimum interpolation method is the cubic interpolation. The interested reader can find detailed analysis given by McFadden [17].

The resampling requires the rotational speed measurement on a shaft. When a mechanical system consists of many shafts, as in the wind turbine drivetrain, it is important to decide on which shaft the speed should be measured. Since the goal is to have high order resolution, it is advantageous to measure the rotational speed of the fastest shaft which is the generator shaft. All shafts have locked frequencies, so all the other shafts in the system can be easily located on the order spectrum.

2.6 Time Synchronous Averaging

Time Synchronous Averaging (TSA) is a technique for detection of gear related faults. The interested reader can find the detailed description given by Wang et al. [18] with an application for a gear tooth crack by Wang [19]. It is based on obtaining a signal describing one revolution of the considered gear by averaging the vibration signal over a number of revolutions. The resulting signal represents the meshing of the teeth over one complete revolution, where the pattern of tooth meshing becomes visible and it is possible to identify e.g. a cracked tooth.

In practical applications sampling is performed with a constant frequency and the machine often contains several gears, such as in the case of a wind turbine. Since averaging requires an integer number of samples per one gear revolution, if the rotational speed is not perfectly constant the vibration signal must first be resampled, as presented in the previous section. When *y(t)* is the measured and resampled time signal, N is the number of gear revolutions in the signal, L is the number of samples in one revolution and T_s is the resampling period, then the time synchronous averaged signal y_{TSA}, according to Wang and McFadden [20] is defined as:

$$y_{TSA}(t) = \frac{1}{N}\sum_{i=0}^{N-1} y(t + iLT_S) \tag{2.19}$$

The application of TSA is equivalent to applying a comb filter to the vibration signal. Parameter N decides about the selectivity of the filter. In the case of processing signals acquired by online monitoring systems, the length of the acquisition cannot be as large as in the case of laboratory experiments or intermittent offline measurements. This was a strong limitation of the presented approach. The signal averaging improves the Signal-to-Noise Ratio (SNR). After averaging N periods, the SNR improvement ratio is $\sqrt{N}$.

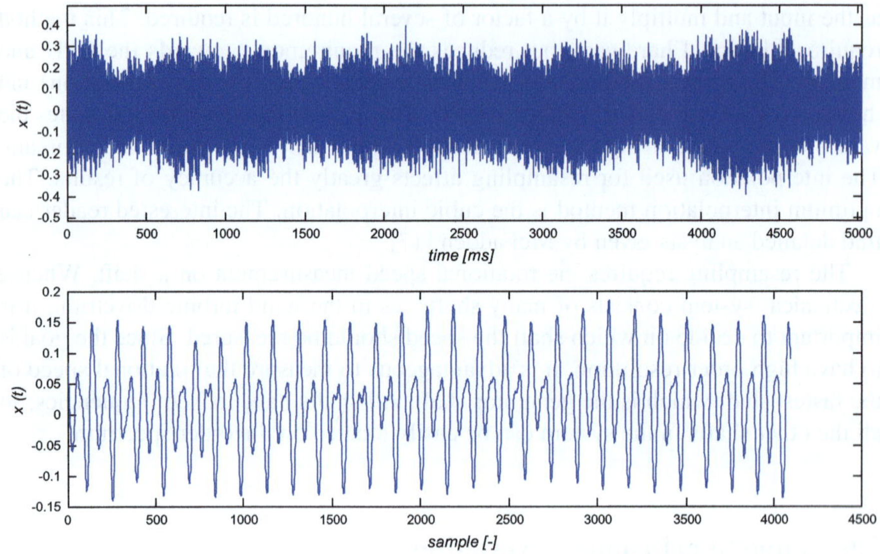

Fig. 2.30 Vibration signal from a wind turbine (top) and result of the TSA algorithm (bottom), resampled to 4096 samples per revolution before averaging. Due to high number of averaged revolutions (841) there is almost no noise in the averaged signal (reprinted from [22] with permission from Elsevier)

The TSA signal tuned to a single revolution of a gear presents meshing over one gear revolution, but in many cases variations in meshing (in particular with varying load) influence the meshing so strongly, that the information about the failure is hidden. To extract possible hidden failures, McFadden and Smith [21] proposed demodulation of the amplitude and phase of the averaged vibration signal. The information related to a tooth failure is then visible in amplitude and phase signals, a typical case being a pulse on the demodulated phase.

The Fig. 2.30 presents the vibration signal from a wind turbine (top). The structure of the signal is very rich and no particular signal structure can be seen. The bottom part of the figure presents the result of the TSA algorithm. The signal was first resampled to 4096 samples per revolution and then averaged over 841 revolutions of the fast speed shaft (i.e. generator shaft). Due to the high number of averaged revolutions, the noise suppression is very good. The obtained waveform reveals 28 clear periods which corresponds with the fact that the fast shaft has 28 teeth on the pinion.

It is important that the same method can be used for other shafts in the drivetrain. The acquired vibration signals collect the vibration from all the shafts and gears in the system. Depending on the location of the sensor the transmission path is very different, but we can assume such a general view at the moment. The averaging period can be assumed equal for a particular shaft rotation period and the vibration from the other shafts will be treated as an unwanted noise and suppressed. The higher the rotational speed of the shaft, the better the noise reduction.

2.7 Time-Frequency Analysis

The analysis methods presented in the previous part of this chapter focused either on the time waveform or the frequency content of the signal. Frequency analysis is a very powerful tool, but assumes stationarity of the analyzed signal. If the signal is non-stationary, as it is in the case of a varying speed wind turbine, the frequency spectrum averages the frequency components in the signal and the time related information is lost. The order analysis is a valuable method of analysis of such signals as presented in the Sect. 2.5. But again, it is focused only on the periodic signal components and additionally requires the phase marker signal to perform the resampling algorithm.

Real life vibration signals consist of many components of varying nature. Apart from periodic sinusoidal components, there are impacts—periodic or non-periodic—with a much wider frequency range. There are also components of varying frequency, e.g. related to the rotational speed, and components of constant frequency, e.g. generated by electric equipment. When one needs to analyze a signal with such a rich characteristic, it is useful to apply a method which combines presentation in both time and frequency domains. This group of methods are *Time-Frequency Analyses*. Many methods have been developed in this area. The most fundamental one is the STFT (Short Time Fourier Transform) which will be described in this chapter. The field of joint time and frequency is much broader. Other analyses are worth mentioning: wavelet transform and Vigner-Wille transform. The interested reader can refer to Flandrin [23].

The short time Fourier transform is defined as:

$$STFT(t, f) = \int_{-\infty}^{\infty} x(\tau)w(\tau - t)e^{-j2\pi f\tau} d\tau \quad (2.20)$$

where:

$x(\tau)$ is the time signal
$w(\tau - t)$ is the windowing function.

The windowing function is a fixed time function of length τ. STFT can be described as a method which assumes stationarity of a signal $x(t)$ over a period of τ. Then, the signal is divided into a series of short (windowed) signals and the Fourier transform is performed for each of the windowed signals. The result of the STFT is a complex matrix which contains both real and imaginary part of the Fourier transform. The matrix has two dimensions, namely time and frequency. For presentation the *spectrogram* is used, which is squared magnitude of STFT and is given by (2.21):

$$SP(t, f) = |STFT(t, f)|^2 \quad (2.21)$$

The chosen window $w(\tau)$ has decisive influence on the STFT. Depending on the window length, the user can choose the resolution in time and frequency domains.

The analysis on a short time window will have high resolution in the time domain, but poor resolution in the frequency domain. The selection of long window will have an opposite effect. In STFT, both domains are related, according to the uncertainty principle. The duration of a signal and its bandwidth follow the condition:

$$B \cdot T \geq \frac{1}{2} \tag{2.22}$$

where:

B is the signal bandwidth
T is the signal duration.

Let us reconsider the signal from the Sect. 2.1, presented in Fig. 2.5. This is the simulated signal which combines vibrations from two shafts at 9 and 24 Hz, the gearbox with two GMF harmonics at 255 and 450 Hz and a component from a faulty rolling element bearing, i.e. 124 impulses/s with carrier frequency around 4 kHz. The sampling frequency is 25 kHz. The Fig. 2.31 presents the spectrogram with 32 frequency bins, so one bin is 781 Hz. Repetitive impulses around 4 kHz can be seen quite clearly. The shaft and GMF frequencies are not possible to locate, as the frequency bin is much larger than these frequencies. The Fig. 2.32 presents the spectrogram with 512 frequency bins, so one bin is now 24.4 Hz. Repetitive impulses around 4 kHz cannot be seen now, but a wideband signal at that frequency is visible. GMF harmonics are now possible to locate. The shaft frequencies have frequency lower than the bin (24.4 Hz) and cannot be distinguished.

A spectrogram is useful for analyzing complex signals when machine operating conditions are changing. Typical cases are transient conditions, e.g. run-ups or coast-downs of a turbine. The Fig. 2.33 presents the spectrogram of a signal from the Fig. 2.8. The signal was acquired during the stop of a wind turbine. The signal is very complex, consisting of decaying vibration and several impulses. All the component frequencies decrease as well making the signal analysis much harder. On the spectrogram plot several components can be identified. First of all, there are no clear decreasing harmonic lines, thus the decrease in the rotational frequency is limited. There are three frequency bands in which the signal energy is concentrated. The highest one is around 3300 Hz, the second one around 1200 Hz and the lowest one around 400 Hz. The amplitude of all three components decreases. Additionally, one can spot vertical lines (around 0.25, 0.5, 0.75, 1.05 s and so on until 3.8 s). These phenomena are short in time and wide in frequency being short impulses. The spacing between them increases with time, due to the slowing down of the turbine.

Please note that all the spectrograms have been converted to the greyscale colormap. This creates clearer pictures in black and white print, but in practice often colormaps are used as it provides better readability.

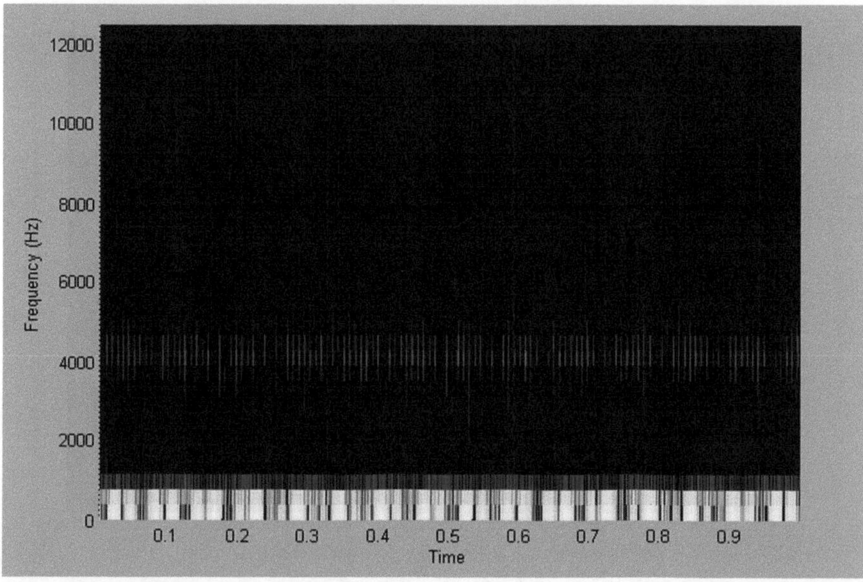

Fig. 2.31 Spectrogram of a signal from Fig. 2.5 obtained for 32 frequency bins. Repetitive impulses around 4 kHz can be seen, but shaft and GMF frequencies are not possible to locate

2.8 Vibration Signal Features

Vibration-based condition monitoring of wind turbines is a process of assessing the technical conditions of mechanical elements of a drivetrain. It uses the analysis of vibration signatures recorded in different locations. In practice, sensors are located (and typically also named) correspondingly to mechanical elements, i.e. housing of the main bearing, planetary gearbox, parallel gearbox, and generator's bearings. In this way, a rise of any health indicator (also called a "diagnostic indicator" or a "feature") is expected to indicate a fault of the element on which the sensor is mounted. For instance, a rise of any vibration-based indicator calculated from a signal recorded from housing of a generator bearing theoretically indicates a fault of this bearing.

However, this assumption is not always valid, because a working bearing is not an autonomous system, but it is mechanically connected/coupled and therefore, from vibrations point-of-view, influenced by other mechanical elements, e.g. a shaft, a motor or a frame. As a result, a sensor mounted at the housing of a bearing might contain signatures of faults not directly related to the technical condition of a bearing, namely its races, cage and rolling elements.

For this reason, commercial systems do not rely solely on attractive indicators which are able to identify a particular cause of mechanical malfunction, but include general indicators as well. The role of those general indicators is to indicate the fact that a machine experiences a fault without any detailed information about the

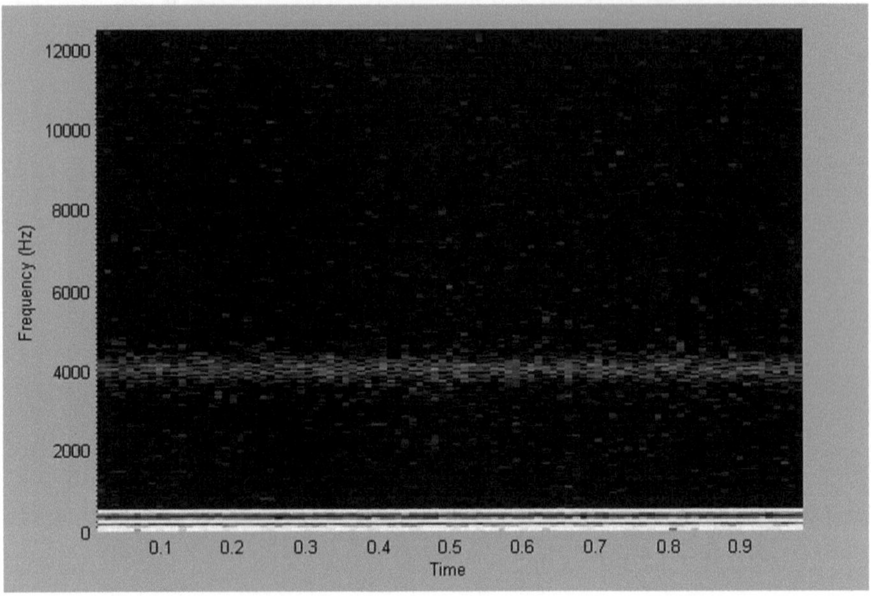

Fig. 2.32 Spectrogram of a signal from Fig. 2.5 obtained for 512 frequency bins. Repetitive nature of impulses around 4 kHz cannot be seen, but a wideband signal at that frequency is visible. GMF harmonics are now possible to locate. The shaft frequencies have value lower than the bin (24.4 Hz) and cannot be distinguished

mechanical element that causes excessive vibrations. Such general features are called "broadband" features because, from definition, they are calculated without any filters, i.e. they are calculated taking into account all signal components, from a large (or "broad") frequency band. Due to some practical considerations, a band for many analyses is defined, but it is much wider than for features monitoring a particular component.

2.8.1 *Broadband Features*

In consequence, there are two basic groups of analyses: broadband, and narrowband, i.e. based on selected spectral lines. Basic broadband analysis parameters are:

- root mean square,
- peak value,
- crest factor,
- kurtosis.

Root Mean Square (RMS) is a very popular signal feature describing the energy of the signal x (2.23):

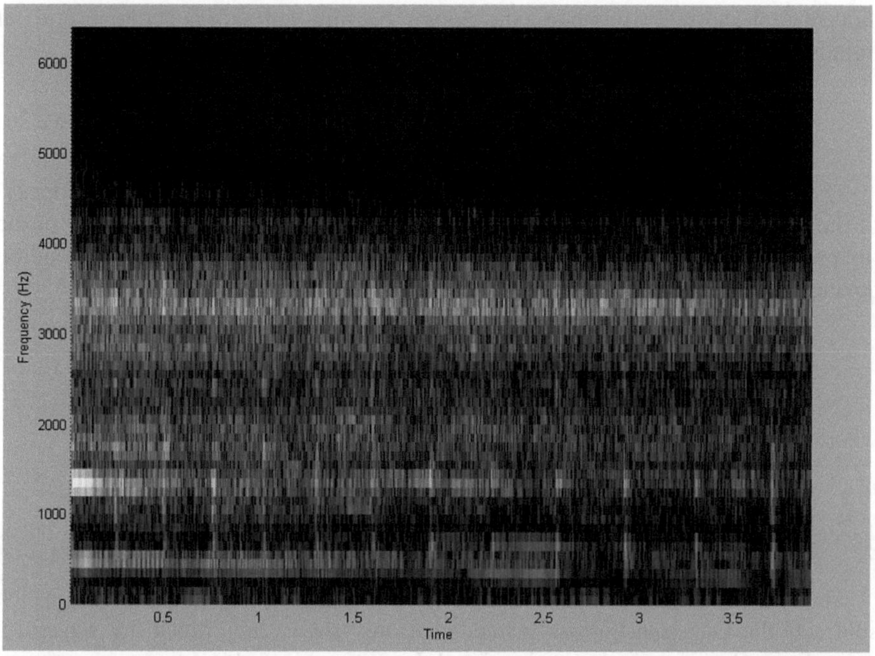

Fig. 2.33 Spectrogram of a signal from Fig. 2.8 obtained for 512 frequency bins. Several signal components can be identified, including three principal bands and repetitive impulses of a decreasing period

$$RMS = \sqrt{E\left(x^2\right)} \tag{2.23}$$

where:

E mean value operator

Peak value (also referred to as peak-peak) informs about maximum peaks in the signal (2.24):

$$PP = x_{\max} - x_{\min} \tag{2.24}$$

Quite often the signature zero-peak is used, as it is easier to compare with particular signal components (2.25):

$$ZP = \frac{x_{\max} - x_{\min}}{2} \tag{2.25}$$

The crest factor is the ratio of peak amplitude of the signal to its rms value. The reason to use this feature was to detect peaks in signals generated by machinery of changing operating point. Such a change modifies generated vibration, together with rms and peak values. Since these features are correlated, their ratio should be load independent and will rise when a developing bearing fault will generate a series of impulses. The crest factor (2.26) is thus:

$$CF = \frac{ZP}{RMS} \tag{2.26}$$

The last broadband feature is the kurtosis. This value represents the "peakedness" of the signal. It is used in statistical analysis and is the 4th central moment of the time series x (2.27). It is more sensitive to peaks in the signal than peak or standard deviation due to the 4th power in the formula:

$$K = \frac{\int_{-\infty}^{\infty} (x - \mu)^4 p(x)dx}{\sigma^4} \tag{2.27}$$

where:

μ mean value of x,
$p(x)$ probability density of x,
σ standard deviation of x.

The kurtosis for the Gaussian random series equals 3. There is a definition which subtracts the Gaussian kurtosis value, so it only shows how much the waveform exceeds the Gaussian random noise signal. In some sources this estimator is referred to as Excess Kurtosis (2.28):

$$EK = \frac{\int_{-\infty}^{\infty} (x - \mu)^4 p(x)dx}{\sigma^4} - 3 \tag{2.28}$$

The most commonly used broadband signal features are the root mean square and the zero peak. The rms vibration levels are defined in the norms, e.g. ISO 10816, which will be discussed in the Sect. 2.9. For relative vibration where rub is the important risk, the peak-peak feature is used. This is a common case in analysis of shaft displacement vibration signals, used in large turbomachinery with sliding bearings. This type of measurements is described in the norm ISO 7919, but it is not used for wind turbines and will not be discussed here. For wind turbines, peak value indicates presence of peaks in the signal and can be a symptom of REB or gear faults.

2.8.2 *Narrowband Features*

The narrowband signal features present a measure describing a particular component linked with a characteristic frequency. Then analysis techniques based on selected spectral lines reflect particular frequencies generated by a certain component. Such frequencies are e.g.: gear mesh, low shaft harmonics, or bearing harmonics. Since such characteristic frequencies are well known, it is sufficient to monitor only a narrow band around the characteristic frequency. In most cases, the energy of the signal in the narrow band is calculated. Such an operation can be performed in the frequency domain, after the spectrum is calculated or in the time domain, when the output of a narrowband filter yields the sought result. The first approach requires spectrum calculation, but when many dozen narrowband estimators are calculated continuously online (as is the case in modern CMS), it is more popular implementation method. There are systems in which, instead of the energy in the band, the peak value in the band is returned as the narrow band estimator.

Due to the different nature of events and to variations in the rotational speed, four basic types of spectra may be investigated for the calculation of narrowband features:

- frequency spectrum,
- order spectrum,
- envelope spectrum,
- envelope order spectrum.

The second and fourth spectra use the spectrum of the resampled signal and are preferred for the machines with varying rotational speed. The third and fourth spectra are used when the envelope spectrum should be investigated, which is predominantly used for the rolling element bearing condition monitoring.

Certainly, several more advanced analyses have been developed, but those presented above are typically implemented in commercial online monitoring and diagnostic systems. This is caused by the fact that those analyses are easy to understand by a majority of vibration experts who in the main are rather practice than theory oriented. On the other hand, the required system configuration may be laborious, but it is not complex.

Both broadband and narrowband features are used for machine monitoring. The examples of analyses applied for a wind turbine are presented in the Table 2.2. The table covers the features monitoring the general condition of the wind turbine as well as those monitoring particular components. General condition is monitored with broadband features, thus it can be calculated from the raw time signal. All the frequencies related to rotating elements should be calculated from the order spectrum. Additionally, the REB elements are best monitored on the resampled enveloped spectrum. All the structural components can be monitored on the "normal" frequency spectrum without resampling, as these frequencies do not change proportionally to the rotational speed. For a real wind turbine, the full list of characteristic frequencies depends on complexity of the drivetrain and for all the mechanical components often exceeds 150.

One needs to remember that monitoring of such an object as a wind turbine must cope with high variation of its vibration behavior, as discussed in the previous chapter. This is primarily caused by frequent changes in wind speed, which in turn cause changes in the output power and—to a lesser extent—in the rotational speed. Since vibration levels depend heavily on those process parameters, it is necessary to define several states. A defined state most often defines a range, e.g. generator

Table 2.2 Standard analysis techniques of wind turbine signals

Monitored component	Analysis	Base
General condition	Rms, zp, crest, kurtosis	Time signal
Shafts	Rotational freqs 1X, 2X	Order spectrum
Main rotor	Blade pass freqs 1X, 2X	Order spectrum
Gears	Mesh freqs 1X, 2X, 3X, sidebands	Order spectrum
Bearings	REB freqs (BPFI, BPFO, BSF, FTF)	Envelope order spectrum
Tower resonance	Structural resonances	Freq. spectrum

output power above 80% of the nominal one. Only values from the same state should be compared. Also, alarm limit values are defined for a given state and can only be analyzed when turbine is operating in that state.

2.9 Vibration-Based Norms and Guidelines

2.9.1 Overview

A wind turbine takes advantage of the kinetic energy of the wind and transfers it into electrical energy. As an energy-source object, a wind turbine is classified as a renewable source of energy. On the other hand, as a mechanical object, a wind turbine is classified as a machine working under highly non-stationary operational parameters. For this reason, in contrary to other machinery like steam turbines, a wind turbine is considered to be a more challenging machine in terms of vibration-based diagnostics.

After the first few years of development of the condition monitoring technology, several solutions were offered to the wind turbine owners and operators. These CMS solutions varied greatly regarding the data acquisition solutions, applied signal processing methods, data storage policies as well as data presentation options. The market required an independent body to investigate the problem of wind turbine condition monitoring and agree on the set of methods which are compulsory for a CMS in the wind turbine industry. An important role was played by insurance companies, as they have a clear incentive for innovative risk reduction technologies. First, several insurance companies proposed guidelines which should be followed by CMS manufacturers. In parallel, discussions and actions were initiated to prepare a formalized set of methods for wind turbines fault detection and localization. This led to a new variant of the popular ISO 10816 norm, dedicated for wind turbines.

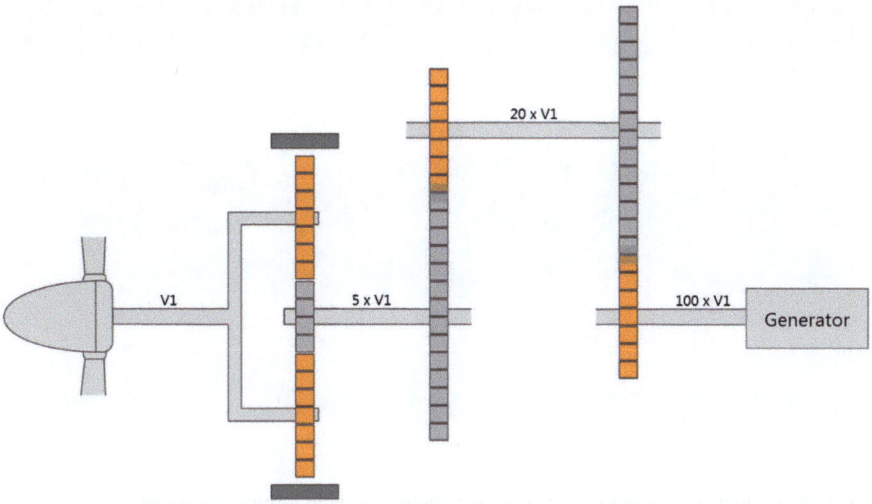

Fig. 2.34 Approximate relative speed ratios in a wind turbine

Let us once again recall specific features which make monitoring of wind turbines harder that of other "standard" machines. The Fig. 2.34 illustrates how a rotational movement of the main rotor blades is transferred by a set of gears into the generator. Approximate relative speed ratios are presented. Because the blades operate with variable speed and torque following the wind, the assessment of the technical conditions of a wind turbine inherently sets additional requirements. These requirements generally include:

- a need to analyze relatively long signals, even hundreds of seconds, due to very slow rotational speed of the main rotor,
- a need to use sensors of relatively high passbands, up to 10 kHz, due to high-frequency modulating components,
- a need to use high resolution ADC converters due to relatively large difference between the speed of the rotor and the speed of the generator, and thus low energy of slow rolling elements vibrations,
- a need to register many process parameters due to essential influence on the vibration levels, especially rotational speed and generator power (an exemplary relationship between the process parameters (power and speed) and vibration waveform is illustrated in the Fig. 2.35),
- a need to define user operational states for an effective selection of registered vibrations,
- a need to automate monitoring of entire windfarms, as the cost of human labor must be minimized.

Over the years, classical diagnostic approaches developed for stationary machines have been used to assess the technical condition of wind turbines, but in many cases,

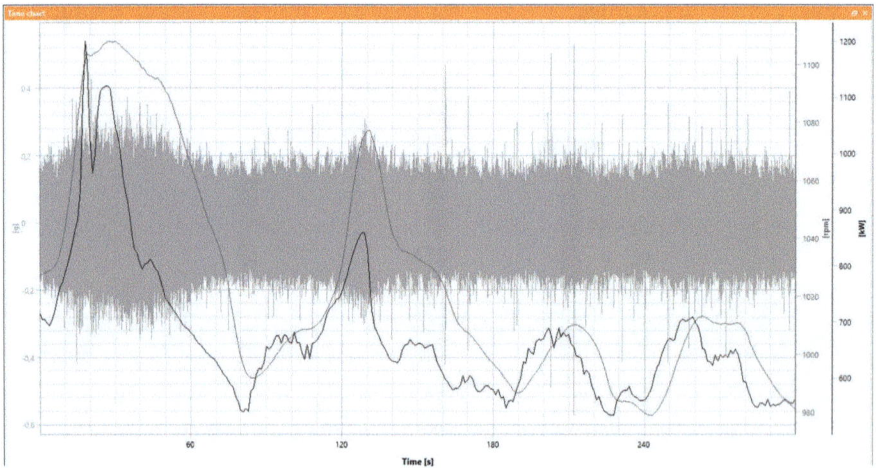

Fig. 2.35 Influence of variation of process parameters on wind turbine vibrations

catastrophic failures continue to occur. These events increased interest in the topic by insurance companies and in 2003 the requirements of vibration monitoring systems for wind turbines were first published by Gellermann and Walter [24] from Allianz Zentrum fur Technik. It is a set of requirements which should be fulfilled by a system to obtain the insurer's recommendation to be installed on a wind turbine. It was as late as the first decade of the XXI century when formal requirements concerning guidelines for configuration, installation, and functions of monitoring systems dedicated to wind turbines became available. The most common guidelines were issued by Allianz, Gellermann and Walter [24] and Germanisher Lloyd in a series of documents (GL [25–27]). Then, the important set of requirements was published by Verein Deutscher Ingenieure (The Association of German Engineers), (VDI [28, 29]). Finally, the first international standard dedicated to wind turbines was published by ISO [30].

New, more up-to-date versions of these documents are elaborated and published. One should realize that they share a lot of common information. In order to present the most important facts, yet to limit the length of the chapter, only selected topics will be presented here. More attention will be given to the ISO 10816-21 standard. It is worth mentioning that research works aiming at classification of methods of diagnostics of wind turbines are constantly in progress. A summary of some recent works might be found in works by Fischer and Coronado [31]. There are also active research groups, which continue to study various aspects of wind turbine fault scenarios and mechanisms. The example of such a group is Gearbox Reliability Collaborative organized by the National Renewable Energy Laboratory at Boulder, Colorado in the US. Several results of the research and the tests are available on the NREL website—see (NREL [32]).

2.9.2 Allianz Guidelines

Allianz Zentrum fuer Technik (AZT) published its first set of guidelines for wind turbines monitoring in 2003, see Gellermann and Walter [24]. In the same year the first version of guidelines was published by Germanisher Lloyd [25] which was then updated a few years later (see GL [26, 27]). Since both, AZT and GL share a great deal of similar approach, only AZT will be discussed here with a few concluding remarks at the end of the chapter about the GL approach.

The AZT guideline is a comprehensive set of requirements applying to all the parts of the condition monitoring system. The document lists the difficulties regarding the performance of WT CMS, namely: dynamic operation behavior, multitude of internal excitations and external vibration excitations. Allianz guidelines state that it is required to cover with the monitoring the main drivetrain components, i.e.:

- main bearing,
- rotor shaft,
- gearing,
- generator,
- tower.

Additionally to the drive train, other wind turbine elements can be monitored optionally. These optional components are: rotor blades, pitch control system, blade bearings, and yaw drives. For key components, the locations and orientation of vibration sensors is defined. Basically, 8 vibration sensors are sufficient for monitoring of all these components. The document also defines frequency ranges for the monitored equipment, as well as the resolution of the A/D converters. The next part lists the characteristic frequencies related to shafts, gearboxes and bearings, which include:

- first shaft order,
- second shaft order,
- blade pass frequencies,
- gear meshing frequencies together with their sidebands,
- all parts of Rolling Element Bearings (BPFO, BPFI, BSF, FTF),
- resonance frequencies of the tower and rotor.

Additionally, the guideline recommends limitation of maximum speed variation during sampling, order tracking and signal preprocessing, and finally a possibility to predefine the number of trend points and raw time signals per unit time. Next, the guidelines tackle far more advanced topics regarding CMS configuration and operation, namely:

- definition of minimum two operation states which take into account speed, power, and wind velocity,
- possible adaptation of threshold values,
- statistical evaluation of health indicators,
- modeling of vibrations for the purpose of diagnostics.

The document recalls diagnostic figures, i.e. the amplitude spectrum and the envelope spectrum as well the notion of "characteristic frequency" implying either a definition of narrowband health indicators or visual spectra analysis for gearbox and REB diagnostics. Additionally, the notion of vibration vector (phase and amplitude) is listed for optional condition monitoring of the tower and nacelle.

In order to analyze the technical state and its evolution, the guideline defines which data should be stored permanently, how often and in which circumstances. The data describing an exceeded alarm threshold should allow for detailed analysis of the root cause of the fault. The CMS should also offer a rich set of graphical tools. Available plots should not only present the data and its analysis (e.g. spectra), but also should facilitate navigation in the system. Such navigation is, for example, required in the case of exceeded limit when full information should be available quickly.

The guidelines from Germanisher Lloyd, apart from defining requirements towards the CMS itself, also put strong pressure on the center which is supposed to collect the data from hundreds or thousands CMS installed on turbines. Such a center should:

- have access to complete documentation of wind turbines with all information about characteristic frequencies of monitored components,
- continuously monitor all trend values and detect threshold violations,
- have good communication with the wind farms,
- update configuration (incl. alarm levels) after repairs,
- store reports together with the base data.

An important requirement of the document is the formal description of responsibilities of all involved actors. The analysts should receive proper technical training and gain relevant experience. The whole monitoring process, from individual CMS system types up to the diagnostic center, should be periodically re-certified. Thus, compliance to the guidelines will constitute the coherent system for the maintenance of wind farms.

2.9.3 VDI 3834 Guidelines

The VDI 3834 guideline was first published in 2009—see (VDI [28]). It filled the gap which existed for wind turbines, as the standards existing at that time explicitly excluded wind turbines from the field of application. The main goal of the document was to introduce a "recognized code of practice which provides criteria and recommendations regarding the measurement and evaluation of the mechanical vibration of WTs and their components". The guideline introduced 4 groups of wind turbines by location (water/land) and the presence of gearboxes:

- Group 1—onshore wind turbines with gearboxes,
- Group 2—gearless onshore wind turbines,

- Group 3—offshore wind turbines with gearboxes,
- Group 4—gearless offshore wind turbines.

The document applies to turbines of group 1 with a nominal power greater than 100 kW, a steel and concrete tower, with horizontal shaft and several blades.

For each component, the frequency range should be defined in such a way, that the characteristic vibration behavior of the component can be grasped. For example, for the aerodynamically excited vibrations of tower and nacelle, the range of 0.1–10 Hz should be selected. As discussed previously, the key problem with assessment of the wind turbines is rapid and chaotic variation of its operational state. To limit the influence of these variations, the evaluation vibration acceleration should be calculated according to the formula (2.29).

$$a_{\mathrm{rms}\;w0} = \sqrt{\frac{1}{T_0}\int_0^{T_0} a_w^2(t)dt} \tag{2.29}$$

The basic value for evaluation is the frequency limited value of a_w, which is acceleration of a measured signal. The overall estimator is then energy-equivalent and averaged over a period of time. Another estimator introduced by the guideline is the evaluation vibration velocity v_{w0}, which is calculated in the same way. An important parameter is T_0 which defines how long the evaluation period is. For the lowest frequencies, e.g. tower and nacelle, presented above, T_0 should be 10 min long.

The guideline acknowledges that there may be periods in which it may not be possible to collect the data for the whole T_0 period. In such a case, a division of this period is allowed, according to the formula (2.30)

$$a_{\mathrm{rms}\;w} = \sqrt{\frac{1}{T_0}\sum_{i=1}^{n} a_{\mathrm{rms}\;i}^2 * T_i} \tag{2.30}$$

where: $T_0 = \sum_{i=1}^{n} T_i$

The evaluation period T_0 can be composed from shorter ones: T_e, but these should not be shorter than 1 min.

The guideline defines all the important parameters of the measurement system and its configuration. It is recommended to measure the acceleration in units of "m^2/s" and the vibration velocity in "mm/s". The sensors in the nacelle should be mounted on both sides of the main bearing mounting, and at the end of the drive system on either side of the generator or the main frame. The sensors on the tower should be mounted below the nacelle bearings. The norm also recommends the directions in which the measurements should be taken for every location. The data acquisition and signal processing parts are also described. The key elements are: vibration transducers, amplifiers, signal conditioning (with filters), signal processing unit, the display and the storage equipment. The allowable accuracies of the measurement path are defined, as well as the tolerances of the band pass filters.

It is interesting that the document specifies requirements towards the measurement of relative shaft vibrations which are closely related to the requirements posed by the DIN ISO 7919-1. As it is the fundamental standard for this type of measurements, it is used for the sliding bearings only which are not used in wind turbines.

For diagnostics of the main bearing, the measurement of the RMS of vibrations is carried out on the bearing housing - the signal should be measured in the frequency range of 0.1–10 Hz (the lower limit must be below the main shaft speed). Like in the case of the tower, the averaging period T_0 should last 10 min. For two bearings, measurements are made in the same way independently. For gearbox diagnostics, VDI 3834 recommends measurement of the vibration acceleration value in the frequency range of 0.1–10 Hz (here again, it is stressed out that the lower limit must be below the main shaft speed) and in the range 10–2000 Hz. Additionally, the norm recommends to measure the RMS of velocity of vibrations in mm/s in the range of 10–1000 Hz. For generator diagnostics, since the rotational speeds are quite high, the document recommends measurement of the vibration acceleration in the range of 10–5000 Hz and the RMS of velocity of vibrations in mm/s in the range of 10–1000 Hz.

VDI 3834 recommends that signals of vibration should be recorded for a turbine operating in the range of 20–100% of rated power. It is accepted to delete data recorded in transition states. The document also reminds that a particular machine may experience a structural resonance which may locally increase the vibration levels. In such a case, the frequency range around the resonance should also be taken into account when configuring the system and evaluating the vibration severity.

In any machinery monitoring program, it is very important to define baseline values which are next used as a comparison with the actual values. The reference base in the VDI 3834 was elaborated as a result of analysis of a large group of wind turbines over the period of years of undamaged operation.

In general, VDI 3834 recommends to define three vibration regions, namely Zone I (good state, suitable for continuous operation), Zone II (investigation is recommended) and Zone III (vibrations are dangerous and damage to the turbine may occur). Limits are given for both acceleration and velocity and if there is a difference in the dynamic state between the two units, the more unfavorable state should be considered. The standard proposes quantitative limits for particular health indicators, presented in Tables 2.3 and 2.4.

Apart from the definition of zones, the guideline also proposes how to set limits for the daily operation of a particular wind turbine. These limits are referred to as the WARNING and ALARM thresholds. The limits can be different for different turbines, but it is recommended that after a base value is established, the WARNING should be the base value increased by 25% of the limit between the zones I/II. Setting the ALARM values, despite similarities between various turbines, is design dependent and the VDI 3834 does not provide instructions on defining absolute

Table 2.3 Limits for permissible acceleration in m/s^2 rms (based on (VDI [29])

Frequency band	Low: 0.1–10 Hz		Mid: 10–2000 Hz		High: 10–5000 Hz	
Zone limits	I/II	II/III	I/II	II/III	I/II	II/III
Nacelle with tower	0.3	0.5	–	–	–	–
Rotor with REB	0.3	0.5	–	–	–	–
Gearbox	0.3	0.5	7.5	12.0		
Generator with REB	–	–	–	–	10.0	16.0

Table 2.4 Limits for permissible velocity in mm/s rms (based on (VDI [29])

Frequency band	Low: 0.1–10 Hz		Mid: 10–1000 Hz	
Zone limits	I/II	II/III	I/II	II/III
Nacelle with tower	60	100	–	–
Rotor with REB	–	–	2.0	3.2
Gearbox	–	–	3.5	5.6
Generator with REB	–	–	6.0	10.0

values. As a general recommendation, the ALARM limits should not be higher than 125% of the limit between zones II/III. Moreover, VDI 3834 states that the thresholds of final limits are set together by the manufacturer and the operator of a wind turbine. It is also permitted that various instances of a wind turbine of the same model will be characterized by different vibration limits.

It is important to verify the limit values after the overhaul of a wind turbine. In such a situation, new data should be collected and new limit values should be established according to a new baseline.

In the year 2015, an updated version of VDI 3834 was published VDI [29]. To a large extent, it is consistent with the 2009 version. There are, however, some alterations:

- combination of onshore and offshore turbines into a single group; the only difference now is between the geared and direct drive designs,
- extension of limits to turbines above 3 MW, as more data were collected,
- removal of relative shaft vibration description, as it is not used in the industrial practice,
- information for balancing the main rotor.

Care was also taken to make the VDI 3834 complementary with the newly issued part of DIN ISO 10816-21, referring specifically to wind turbines. As ISO is widely recognized around the world, the appearance of this part was important for the community.

2.9.4 *ISO 10816-21 Standard*

The International Organization for Standardization (ISO) is probably the most renowned worldwide organization which issues technical standards for various industries. It is composed of national organizations called ISO member bodies which are organized into Technical Committees (TC). The subject of vibrations is very important for evaluation of the technical state of machinery and became subject of work for several TCs. The standard referring to the wind turbines monitoring is relatively new and it is ISO 10816, part 21. The overall norm, namely ISO 10816 is "Mechanical vibration—Evaluation of machine vibration by measurements on nonrotating parts" and refers to measurements on casings and bearing housings done by accelerometers and vibration velocity sensors. The norm describes numerous machinery types and so it is divided into several parts. The Table 2.5 presents the list of parts. Please note that the numbering is not continuous.

The topic of vibration measurement and evaluation is very broad and many more standards deal with it. Table 2.6 summarizes the scope of topics covered by the other frequently used vibration related standards which are helpful for vibration based condition monitoring of wind turbines.

Historically, the ISO 10816-21 is based on various standards and guidelines published in previous years. Probably the largest portion was inherited from VDI 3834 presented in the previous chapter. The reader will spot several similarities, first of all in the concept of averaging vibration levels in order to reduce the influence of varying operational conditions. The other similarity is the very same levels of zone boundaries.

As in VDI 3834, the ISO standard defines 4 groups of wind turbines:

- Group 1—onshore wind turbines with a gearbox,
- Group 2—onshore wind turbines without a gearbox,

Table 2.5 Parts of ISO 10816 standard

Part number	Title
Part 1	General guidelines
Part 2	Land-based steam turbines and generators in excess of 50 MW with normal operating speeds of 1500, 1800, 3000 and 3600 r/min
Part 3	Industrial machines with nominal power above 15 kW and nominal speeds between 120 and 15000 r/min when measured in situ
Part 4	Gas turbine sets with fluid-film bearings
Part 5	Machine sets in hydraulic power generating and pumping plants
Part 6	Reciprocating machines with power ratings above 100 kW
Part 7	Rotodynamic pumps for industrial applications, including measurements on rotating shafts
Part 8	Guidelines for vibrations in reciprocating compressor systems
Part 21	Onshore wind turbines with gearbox

Table 2.6 Other vibration-based norms related to wind turbines

Norm number	Scope
ISO 2041	"Mechanical vibration, shock and condition monitoring—Vocabulary"; Definitions of terms related to monitoring of mechanical vibrations, shocks and condition of machines
ISO 5348	"Mechanical vibration and shock—Mechanical mounting of accelerometers"; Guidelines for mounting methods of accelerometers and influence on measurement results
ISO 13373-2	"Condition monitoring and diagnostics of machines—Vibration condition monitoring—Part 2: Processing, analysis and presentation of vibration data."; Recommendations concerning the processing and presentation procedures of vibration measurement data and vibration indicator analysis for monitoring the state of rotating machine elements
ISO 13379	"Condition monitoring and diagnostics of machines—General guidelines on data interpretation and diagnostics techniques"; Procedures for the implementation of diagnostic methods to facilitate subsequent analyses
ISO 17359	"Condition monitoring and diagnostics of machines—General guidelines"; Guidelines for general procedures to be taken into account when configuring the machine health monitoring system, including references to related standards required in this process
ISO 19201	"Mechanical vibration—Methodology for selecting appropriate machinery vibration standards"; Guidelines for the selection of appropriate vibration measurement and evaluation methods depending on the type of a machine

- Group 3—offshore wind turbines with a gearbox,
- Group 4—offshore wind turbines without a gearbox.

Part 21 of the standard refers only to the group 1. Other groups will be defined when sufficient data is selected. Since the new edition of VDI was published in 2015, there is high probability that it had and it will have some influence on the development of ISO 10816. As for now (2018), the ISO 10816-21 standard refers to wind turbines which share following features:

- are installed onshore
- tower and foundation are made of steel and concrete
- have horizontal axis with several rotor blades
- generator is coupled to the rotor by a gearbox
- generator can be synchronous or asynchronous
- generator is connected to the grid directly or with a converter
- are pitch or stall controlled.

The document provides only general guidelines for evaluating vibration levels in the stable operating range, based on overall (rms) values of acceleration and velocity of vibrations. It does not define any frequency—based characteristic features. Therefore, the standard does not support users in an early phase of fault detection, nor in the assessment of the technical state of rolling element bearings and gears. It is likely that with the development of monitoring techniques and gained experience future editions of the standard will include at least some of these aspects.

The ISO standard stresses the varying operating conditions and their influence on the measured vibration levels. In order to reduce this influence the energy—equivalent averaging is proposed. The exact formulas are the same as proposed in the VDI [see Eqs. (2.29) and (2.30)]. Key parameters for calculation of a_{rmsw} parameters are: evaluation period T_0 and the operating frequency ranges. These values should be presented every time when vibration values are reported.

During the averaging period the generator output power should not be less than 20% of the nominal power. It is allowed not to take into account the vibration levels taken during transient operating conditions, i.e. starting or braking and such results can be excluded from the evaluation. On the other hand, even if beyond the required 20% nominal power range, there are particular conditions (most often output power), when increased vibrations occur. It should also be taken into account when performing measurement and evaluation.

A large part of the standard defines signal features (called characteristic quantities), measuring positions, measuring directions, frequency ranges and averaging periods for key turbine components, i.e. nacelle and tower, rotor main bearing, gearbox and generator.

The authors stress the importance of the proper vibration measurement systems. In general, they should follow requirements posed by the standard ISO 10817-1 [33]. The most important requirements are the measurement uncertainty which should not exceed ±10% and the filter frequency ranges tolerance which should not exceed ±3 dB. It is possible to apply relative shaft vibration measurements, though it is very unpopular. Vibration measurements can be executed with online or offline measurement systems. Exact requirements towards vibration transducers mounted on the machine parts are also presented. It is stressed that for the accurate measurement of a high frequency band the most rigid coupling possible should be applied.

The ISO 10816-21 introduces the same severity zones as VDI, namely Zone I (good), Zone II (investigation recommended) and Zone III (dangerous). Next, guide values for boundaries between the zones are defined. The guide values have exactly the same values as in VDI (see Tables 2.3 and 2.4), though they are organized in a different way. Similarly to VDI, the procedure to define vibration limits for operation is proposed. These limits are called WARNING and ALARM and are defined as in VDI 3834. The authors of the ISO standard recognize that proposed zones and acceptable limits are for evaluation of general state only and do not provide early fault detection. For such purposes, frequency selective characteristic values should be individually defined. The most efficient method for reference values are historical measurements. With such data, baseline values can be defined and then warning and alarm levels can be calculated.

References

1. Randall RB (1987) Frequency analysis. Bruel & Kjaer, Naerum
2. Shin K, Hammond JK (2008) Fundamentals of signal processing for sound and vibration engineers. Wiley, Chichester
3. Braun S (2008) Discover signal processing: an interactive guide for engineers. Wiley Chichester
4. Klein U (2003) Schwingungsdiagnostische Beurteilung von Maschinen und Anlagen. Stahleisen Verlag, Duesseldorf (Vibrodiagnostic assessment of machines and devices, in German)
5. Jardine AK, Lin D, Banjevic D (2006) A review on machinery diagnostics and prognostics implementing condition based maintenance. Mech Syst Sig Process 20(7):1483–1510
6. Siegel D, Zhao W, Lapira E, Abuali M, Lee J (2014) A comparative study on vibration-based condition monitoring algorithms for wind turbine drive trains. Wind Energ 17:695–714
7. Darlow MS, Badgley RH, Hogg GW (1974) Application of high-frequency resonance techniques for bearing diagnostics in helicopter gearboxes. Mechanical Technology Inc., Latham
8. Randall RB, Antoni J, Chobsaard S (2001) The relationship between spectral correlation and envelope analysis for cyclostationary machine signals. Application to ball bearing diagnostics. Mech Syst Sig Process 15(5):945–962
9. Antoni J (2007) Cyclic spectral analysis of rolling-element bearing signals: facts and fictions. J Sound Vib 304:497–529
10. Antoni J (2009) Cyclostationarity by examples. Mech Syst Sig Process 23(4):987–1036
11. Ho D, Randall RB (2000) Optimisation of bearing diagnostic techniques using simulated and actual bearing fault signals. Mech Syst Sig Process 14(5):763–788
12. Randall RB, Antoni J (2011) Rolling element bearing diagnostics—A tutorial. Mech Syst Sig Process 25(2):485–520
13. Barszcz T, Jablonski A (2010) Selected methods of finding optimal center frequency for amplitude demodulation of vibration signals. Diagnostyka 2:25–28
14. Bartelmus W, Zimroz R (2006) Optymalny zakres częstotliwości w procedurze demodulacji amplitudy w zastosowaniu do uszkodzeń lokalnych. Diagnostyka (Optimal frequency range for amplitude demodulation for local fault detection, in Polish) 1(37):141–150
15. Barszcz T, Jablonski A (2011) A novel method for the optimal band selection for vibration signal demodulation and comparison with the Kurtogram. Mech Syst Sig Process 25(1):431–451
16. Stearns S (1996) Signal processing algorithms in Matlab. Prentice Hall Inc/
17. McFadden PD (1989) Interpolation techniques for time domain averaging of gear vibration. Mech Syst Sig Process 3:87–97
18. Wang W, Ismail F, Golnaraghi MF (2001) Assessment of gear damage monitoring techniques using vibration measurements. Mech Syst Sig Process 15:905–922
19. Wang W (2001) Early detection of gear tooth cracking using the resonance demodulation technique. Mech Syst Sig Process 15:887–903
20. Wang WJ, McFadden PD (1995) Decomposition of gear motion signals and its application to gearbox diagnostics. J Vib Acoust 117:363–369
21. McFadden PD, Smith JD (1985) A signal processing technique for detecting local defects in a gear from the signal average of the vibration. P I Mech Eng 199:287–292
22. Barszcz T, Randall RB (2009) Application of spectral kurtosis for detection of a tooth crack in the planetary gear of a wind turbine. Mech Syst Sig Process 23:1352–1365
23. Flandrin P (1998) Time-frequency/Time-scale analysis. Academic Press
24. Gellermann T, Walter G (2003) Anforderungen an Condition Monitoring Systeme für Windenergieanlagen. Allianz Zentrum fur Technik
25. Germanischer Lloyd WindEnergie GmbH (2003) Richtlinie für die Zertifizierung von Condition Monitoring Systemen für Windenergieanlagen
26. Germanischer Lloyd WindEnergie GmbH (2007) Richtlinie für die Zertifizierung von Condition Monitoring Systemen für Windenergieanlagen
27. Germanischer Lloyd WindEnergie GmbH (2013) Richtlinie für die Zertifizierung von Condition Monitoring Systemen für Windenergieanlagen

28. Vereins Deutscher Ingenieure (2009) VDI 3834-1:2009-03. Measurement and evaluation of the mechanical vibration of wind energy turbines and their components—Onshore wind energy turbines with gears. Verlag des Vereins Deutscher Ingenieure
29. Vereins Deutscher Ingenieure (2015) VDI 3834-1:2015-08. Measurement and evaluation of the mechanical vibration of wind energy turbines and their components—Onshore wind energy turbines with gears. Verlag des Vereins Deutscher Ingenieure
30. International Standard Organization (2015) ISO 10816-21:2015. Mechanical vibration—Evaluation of machine vibration by measurements on nonrotating parts—Part 21: Horizontal axis wind turbines with gearbox
31. Fischer K, Coronado D (2015) Condition monitoring of wind turbines: state of the art, user experience and recommendations. VGB PowerTech 7:51–56
32. National Renewable Energy Laboratory (2015) Gearbox reliability collaborative research. https://www.nrel.gov/wind/grc-research.html Accessed on 20 Mar 2018
33. International Standard Organization (1998) ISO 10817-1:1998. Rotating shaft vibration measuring systems—Part 1: Relative and absolute sensing of radial vibration
34. Antoni J (2007) Fast computation of the kurtogram for the detection of transient faults. Mechanical Systems and Signal Processing 21 (1):108–124

Chapter 3
Condition Monitoring Systems

3.1 Sensors

Sensors provide inputs for any further signal processing and decision making. Sensors in condition monitoring are divided into two principal sensor types, namely dynamic and process. The fundamental difference between these two is the sampling rate at which signals are acquired.

For dynamic sensors (in monitoring of wind turbines these are almost only vibration sensors, except research installations) the sampling frequency is in the range of 10–50 kHz, with a typical value of 25 kHz. In some systems frequencies up to 100 kHz are used. Such sampling rates (multiplied by the number of sensors) generate a vast stream of data which needs to be analyzed by the system. Moreover, it is both uneconomical and unnecessary to store all the dynamic data. Instead, signal features (as presented in the Chap. 2) are calculated from vibration signals and only the features are then analyzed and stored. Some systems, though, have the ability to store raw vibration data as well. To save memory, records of raw vibration data several seconds long each are stored only periodically (e.g. once per week) or in case when the event is detected (e.g. alarm threshold violation).

The other type of sensors are process values sensors. The sampling rate in this case is in the range of one second, sometimes much longer. These variables provide context for vibration data and they may be for example rotational speed, output power, oil temperatures, ambient temperature etc. The data from process sensors is used to understand the operational state of the machine (i.e. stopped, running up, partial power, nominal power, etc.). The operating state often impacts the machine vibrations far more than an upcoming fault, as discussed in the Chap. 1.

© Springer Nature Switzerland AG 2019

T. Barszcz, *Vibration-Based Condition Monitoring of Wind Turbines*, Applied Condition Monitoring 14, https://doi.org/10.1007/978-3-030-05971-2_3

3.1.1 Vibration Acceleration Sensors

Mechanical vibration can be measured by three different though interconnected values, namely displacement, velocity or acceleration. It is important to understand the relationships between these values. If there is a vibrating body, in a general case its movement can be described with an arbitrary displacement function *x(t)*. The other two values can be calculated as presented in the Eqs. (3.1) and (3.2).

$$v(t) = \frac{dx(t)}{dt} \tag{3.1}$$

$$a(t) = \frac{d^2x(t)}{dt^2} = \frac{dv(t)}{dt} \tag{3.2}$$

If the measured value is the displacement, velocity can be obtained by the differentiation and the acceleration by second order differentiation. In a similar way the opposite operations can be performed, so if acceleration is given a single integration will yield velocity and a double integration—displacement. When we assume that a vibration displacement signal $x(t)$ is a harmonic function given by (3.3), then velocity of $x(t)$ is given by (3.4) and acceleration by (3.5).

$$x(t) = A_0 \sin \omega t \tag{3.3}$$

$$v(t) = A_0\omega \cos \omega t \tag{3.4}$$

$$a(t) = -A_0\omega^2 \sin \omega t \tag{3.5}$$

Apart from the phase difference, signals from (3.3), (3.4) and (3.5) differ by the factor ω which is the angular frequency (it equals frequency divided by 2π). Therefore, the transition between displacement, velocity and acceleration can be presented in the frequency domain. Differentiation in the time domain is equivalent to multiplication by ω in the frequency domain. On the discrete frequency spectrum (see Chap. 2) differentiation can be achieved by multiplying each spectral line by its angular frequency ω. The important consequence of this fact is that acceleration is the best value to analyze high frequencies, velocity—for middle ones and displacement—for low frequencies. Type of a chosen sensor also depends on mounting constraints, since acceleration sensors are the easiest to mount on rotating machinery. One has to bear in mind that sensor response depends on frequency of a signal and is very small for low frequencies.

As there are three quantities describing mechanical vibration there are also three types of vibration sensors: those which measure vibration displacement, vibration velocity and vibration acceleration accordingly. The third group, i.e. accelerometers are widely used as a data source in wind turbine condition monitoring.

Figure 3.1 presents the internal structure of the accelerometer. This type of a sensor uses the phenomenon of piezoelectric effect. The piezoelectric element is used as a spring which connects the seismic mass and the measured body (the sensor enclosure). When acceleration is applied to the sensor the piezoelectric element

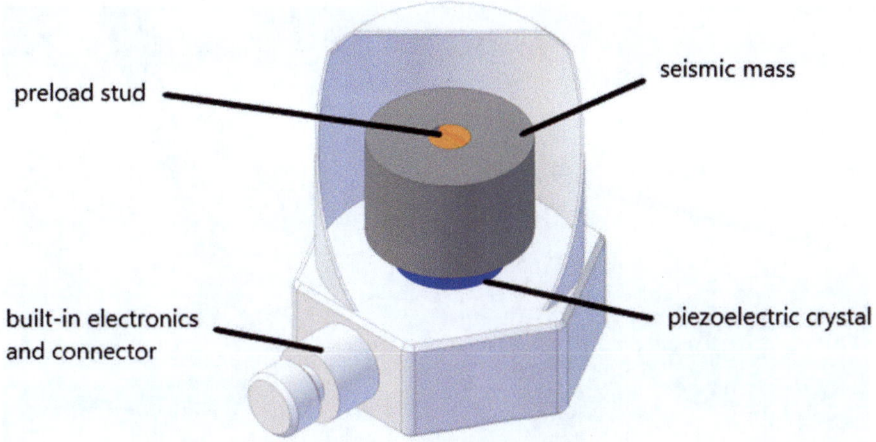

Fig. 3.1 Internal structure of the piezoelectric acceleration sensor

generates the electric charge. The charge is next transformed by the built-in electronics. It serves as a transducer from electric charge to a current signal which is more robust to disturbances on the path between a sensor and an acquisition unit. There are also sensor models in which transducers are separate devices. They are used in special applications, e.g. high temperature environments but they are not used in wind turbine condition monitoring. Figure 3.2 presents an accelerometer mounted to the wind turbine gearbox. The security bracket is mounted around to prevent it from an accidental damage during maintenance works.

According to the norm ISO 10816—21:2015 "Mechanical vibration—Evaluation of machine vibration by measurements on non-rotating parts—Part 21: Horizontal axis wind turbines with gearbox" (see ISO10816-21 [1]), the vibration sensor should cover the following turbine parts:

- nacelle and tower,
- rotor shaft with main bearing,
- gearbox,
- generator.

To limit the number of vibration sensors and measurement channels it is common to use low frequency accelerometers mounted on the main bearing and the planetary gearbox for monitoring the nacelle and the tower. Figure 3.3 presents the example of a vibration sensor location on the drivetrain. One sensor is used for the main bearing, three sensors are used for the gearbox and another two for the generator. Sensors should cover vertical and horizontal orientation. One of the sensors is often mounted in the axial direction. Please note that for particular turbine types this scheme may vary. Sometimes additional accelerometers are used for the nacelle and the tower. Also, for complex gearbox designs more sensors may be used. In most installations 6 vibration sensors are sufficient for monitoring the drivetrain. For complex drivetrains this number can rise up to 14 sensors.

Fig. 3.2 Accelerometer mounted on top of the wind turbine gearbox

The important consideration for the accelerometers used for slowly rotating parts, like the main shaft or the planetary gearbox, is the lowest frequency they can transmit. A typical value is about 2 Hz and it is not sufficient for turbines where the typical main rotor speed is 20 rpm, i.e. 0.33 Hz. For such a purpose special models of accelerometers should be used. Due to the low rotational speed the vibration energy is also small and consequently high sensitivity sensors should be used. Here, the 601A02 accelerometer was selected as an example of such a sensor suited for monitoring a low speed part, as presented in PCB 601A02 [2]. Figure 3.4 presents the external view of this sensor.

Table 3.1 shows its key technical parameters. For other drivetrain components standard accelerometers can be used.

Another type of accelerometers are those based on the Micro Electro Mechanical Systems (MEMS) technology. The MEMS technology allowed to build very small (size of a fraction of a millimeter) sensors, including accelerometers (see Albbarbar and Teay [3]). Important advantages of this technology are: small size, low price and good performance in low frequencies. Some popular fields of use were the first smartphones, hard disks and airbags, as well as dozens of others. The application to machinery condition monitoring was not possible due to too high a noise ratio and too low bandwidth. The bandwidth was in a range of several hundred Hertz up

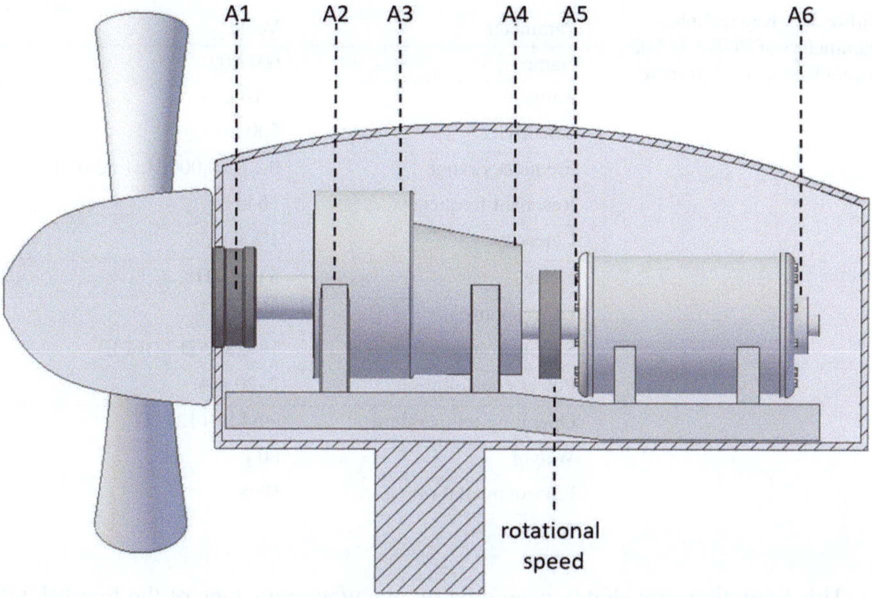

Fig. 3.3 Example locations of vibration sensors

Fig. 3.4 Accelerometer PCB 601A02 used for condition monitoring of slowly rotating parts

to around 1.5 kHz. The noise level is also a limiting factor. As seen in the tables comparing the two technologies the noise level for the MEMS sensor is over 500 times higher.

Table 3.1 Key technical parameters of PCB 601A02 piezoelectric accelerometer

Parameter	Value
Name	601A02
Range	±10 g
Sensitivity	500 mV/g
Frequency range	0.17–10,000 Hz (±3 dB)
Resonant frequency	16 kHz
Linearity	1%
Noise	5 $\mu g/\sqrt{Hz}$
DC response	No
Supply voltage	18–28 V DC (IEPE)
Power consumption	2–20 mA
Operating temperature	−54 to +121 °C
Weight	80 g
Environmental rating	IP68

This limitations are slowly overcome by manufacturers. One of the first MEMS with bandwidth suitable for machinery condition monitoring was ADXL001 by Analog Devices [4]. Figure 3.5 presents the printed circuit board with two ADXL001 sensors to measure vibrations in two directions. Spence and Chong [5] informed about a newly developed MEMS based sensor with the noise level reduced by two ranges of magnitude. The authors wrote about the development of a new condition monitoring system using the MEMS technology at a cost being even 10 times lower than the CMS using traditional vibration sensors. Table 3.2 presents key parameters of the ADXL001 accelerometer. It is clearly seen that the main disadvantages of MEMS accelerometers in comparison to piezoelectric ones are: lower sensitivity and much higher noise level. The latest MEMS accelerometers, e.g. ADXL0002 by Analog Devices [6] have a greatly reduced noise level. This level is now ca 5 times higher than for piezoelectric sensors. The list price, however, is not 10 times lower but rather 2–4 times.

The MEMS integrated circuit cannot be installed on a machine "as is" and needs to be built into a proper enclosure. Nevertheless, the MEMS circuit makes a vital part of the industrial sensor and decides about its key features. There are already industrial grade sensors available on the market offering industrial quality using the MEMS technology, e.g. VSA006 from the German company ifm, see IFM VSA006 [7]. Figure 3.6 presents the external view of the sensor and Table 3.3—key parameters of VSA006 accelerometer. With such a pace in technical development, especially with decreasing noise levels, the MEMS sensors will soon create an increasingly large competitive pressure on the piezoelectric technology.

Fig. 3.5 Printed circuit board with two ADXL001 MEMS sensors (in the bottom)

Table 3.2 Key technical parameters of ADXL001 MEMS accelerometer

Parameter	Value
Name	ADXL001
Range	±70, ±250, ±500 g
Sensitivity	24.2, 6.7, 3.3 mV/g (for 5 V PS)
Resonant frequency	22 kHz
Linearity	0.2% full scale
Noise	Below 2.8 mg/√Hz (for 5 V PS)
DC response	Yes
Supply voltage	3.3–5 V
Power consumption (typ.)	2.5 mA
Operating temperature	−55 to +125 °C
Environmental rating	None

3.1.2 Process Sensors

The second group of sensors are those used to measure slowly changing process values. The most important process sensors are the rotational speed and the output power ones. In wind turbine condition monitoring standard industrial sensors are widely used. The main difference between various sensors is—apart from the measured physical value—the electric standard of the interface to get the data into a CMS. Three of such interfaces will be covered in this chapter, namely impulse, analogue and digital.

The rotational speed sensor is a very important one, not only because it provides crucial information to assess the drivetrain dynamics. The rotational speed, also referred to as a phase marker or tacho, is used in vibration signal acquisition for

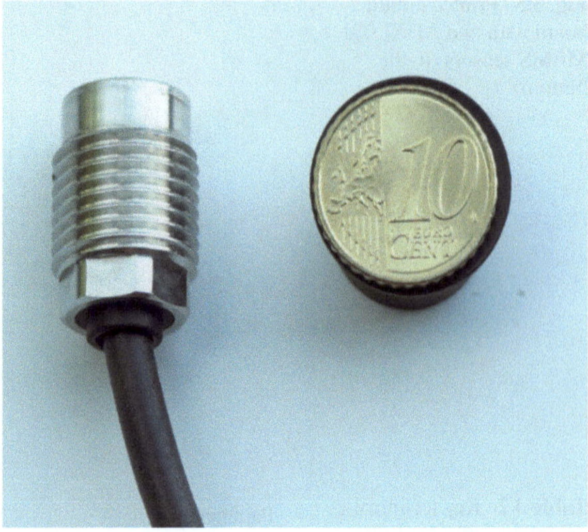

Fig. 3.6 VSA006 MEMS based sensor

Table 3.3 Key technical parameters of VSA006 MEMS accelerometer

Parameter	Value
Name	VSA006
Range	±25 g
Frequency range	0–10 kHz
Linearity	0.2% full scale
Supply voltage	9 V DC
Power consumption (typ.)	<15 mA
Operating temperature	−30 to +85 °C
Weight	253 g
Environmental rating	IP 67

further signal processing. It is used to assess the stability of the shaft rotation and also for the order analysis (i.e. signal resampling), presented in the Sect. 2.5.

Figure 3.7 presents a rotational speed sensor installed on the wind turbine. A wind turbine with a gearbox offers several locations to install a speed sensor, since it has several rotating shafts. In practice, it is optimal to mount the speed sensor on the fastest shaft which is the generator shaft. Next, with a numerical analysis the speed information and knowledge about gear ratios can be used to calculate speed of any other shaft in the drivetrain. The Hall effect sensors are typically used as the tacho sensors due to their robustness and a low cost. For the proper operation, the Hall sensor requires a permanent magnet to be placed on a rotating shaft.

The tacho sensor generates a pulse during every pass of a magnet under the sensor tip. The exact electric parameters of this signal depend on the sensor and the setup of the power supply. In general, key rotational speed signal information is carried with the edges (positive or negative). This type of signal requires an impulse input on the

Fig. 3.7 Rotational speed sensor (Hall sensor) installed on the generator shaft. The permanent magnet installed on the shaft is visible just next to the sensor tip

receiving system side. This is a type of a digital input which can accommodate quick series of pulses and count their number or calculate the number of pulses per a time period (i.e. the rotational speed).

Another group of process sensors are the analogue ones. They are used to measure a variety of physical values in numerous industries. The most popular sensor used in wind turbine condition monitoring is a wind speed and a generator output power sensor. Figure 3.8 presents the picture of the wind sensor installed on a wind turbine. This is a popular sensor where the rotational speed of the rotor with 3 hemispheres is converted into the wind speed. There are also other types of wind speed sensors based on sound speed and Doppler effect measurement. The advantage of such a design is no moving parts.

All the analogue sensors generate an electrical value at its output. The most popular standards are 0–10 V and 4–20 mA. The principle of operation of the standard is that the if the sensor is exposed to the low end of its physical range it generates a low electric value and if it is exposed to the high end—it generates a high electric value. For values in between the sensor generates a value which is linear interpolation of the two boundary ones.

In the recent years more and more popularity has been gained by sensors with digital interfaces, sometimes proprietary to a vendor. This usually happens when the sensor constitutes a part of a larger system, e.g. a turbine control system. In such a situation it is hard to reuse the same sensor and for CMS purposes either a new sensor must be installed or a digital interface to the existing control system can be used.

Fig. 3.8 Wind speed sensor of the mechanical type installed on the top of a nacelle

The third group of sensors are digital ones (also referred to as binary). These sensors report the status of a device which can be in two different states. A typical example is the activation of an electric motor. Digital signals are not often used in WT condition monitoring. One of a few examples can be information about activation of a yaw drive. Operation of this drive generates significant vibrations which are an important disturbance to the vibration analysis. There are CMS which do not perform data acquisition if this drive is active.

3.2 SCADA Systems

Virtually any wind farm has a system for data collection, storage and display to the operators. This type of system is referred to as the Supervisory Control And Data Acquisition (SCADA) system. Their role is to acquire hundreds of process signals from controllers on each turbine on a wind farm. Next, each signal history is stored in a very large buffer from which it can be retrieved even after several years. SCADA systems also provide the information about the wind farm to the operators. These systems work on fast data links, so the control centers are often far away from the actual site. A typical architecture of the SCADA system for several wind farms connected to the management center is shown in Fig. 3.9.

Each WT controller is a data source and is connected to the Wind Farm SCADA server. Farms are monitored from a remote center, often several farms per a single center. With decreasing cost of technology and increasing costs of manpower, there is a trend to organize higher level centers which collect and analyze data from many

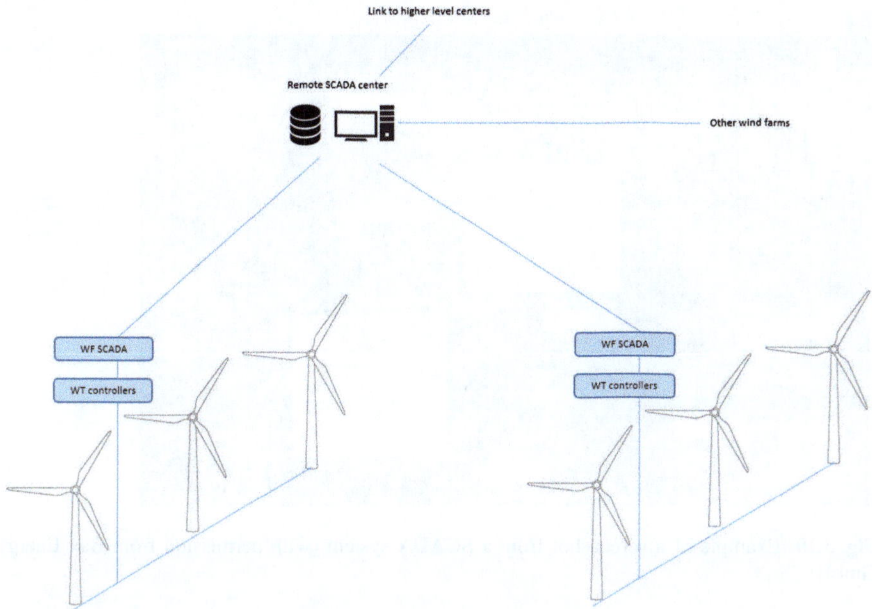

Fig. 3.9 Typical architecture of the SCADA systems connected to a remote center

wind farms. Such centers can be maintained by the manufacturers of SCADA systems. There are also centers maintained by power plant owners and they usually integrate hundreds of wind turbines. In the latter case, wind farm data is stored along with data from other plants, e.g. thermal or hydro plants.

The example of a typical SCADA screen is presented in Fig. 3.10. For a typical SCADA on a wind farm, more than a hundred of parameters are monitored and stored for a single wind turbine. The selected data measured and stored in the system are:

- wind speed,
- wind direction,
- ambient temperature,
- rotational speed of the generator shaft,
- generator active power,
- generator reactive power,
- generator currents,
- generator voltages,
- energy transmitted to the grid,
- pitch angles of blades,
- oil pressures,
- oil temperatures,
- temperatures of bearings (main, generator, gearbox),
- temperatures of generator phase windings,

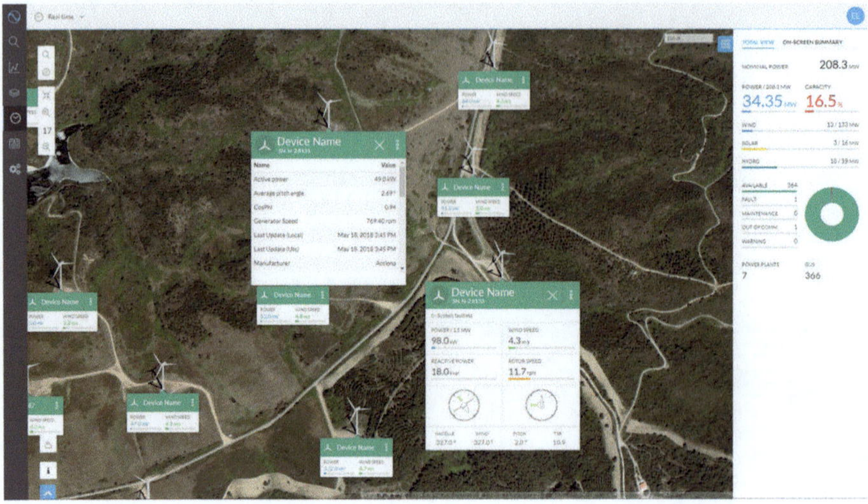

Fig. 3.10 Example of a screenshot from a SCADA system (with permission from Bax Energy GmbH)

- temperatures of transformer windings,
- temperature in nacelle,
- direction of nacelle,
- controller error codes.

Typically, the data are stored every 10 min. Since the data may (and will) vary during this period, the system records the minimum, maximum, average and standard deviation values for each 10 min period. The interested reader can find the details about SCADA systems in several works, e.g. Thomas and McDonald [8]. Communication protocols play a significant role in connecting various SCADA systems with a turbine controller and the CMS. The information about these standards can be found e.g. in Clarke and Reynders [9].

From the perspective of condition monitoring, SCADA technology offers an interesting opportunity. Huge amounts of data are available at hand for analysis. Farm owners do not need to invest in additional condition monitoring equipment, like vibration based CMS. Thus, if a fault can be detected by analyzing already available data, it is a large cost advantage. Even if a fault is detected at a later stage it is still acceptable for some owners. Due to a very rapid development of machine learning methods in the last decade, such an approach has drawn a lot of attention from the scientific community. As a result a large number of papers proposing various health assessment algorithms have been proposed. A detailed review in this field of research can be found in Tautz-Weinert and Watson [10].

With gigabytes and terabytes of collected data, SCADA systems can be used as a platform to implement various health assessment algorithms, as presented in the paragraph above. The gearbox fault detection is a good example of application of data mining techniques. If technical state of a gearbox deteriorates, oil temperature will rise. Oil temperature monitoring is much less sensitive to fault than vibration based methods, but it utilizes the data already present in the SCADA database. Even with such data, fault detection is not an easy endeavor. Figure 3.11 presents the history of a few tags showing one week data. The tags are: active power, ambient temperature, gearbox oil temperature and the wind speed. The variability of the data is very high and has nothing to do with any deterioration of the technical state, but with the changes in wind speed and—to a lesser extent—with the ambient temperature. Once again, it can be observed that the main reason for difficulties in fault detection is the variation of the operational point which needs to be taken care of in the first place.

There are two principal groups of machine learning methods: a supervised (more popular) and an unsupervised one. The supervised learning method requires a large amount of data as an input. Both "healthy" data from a machine without faults and "bad" data (with faults) is needed. For supervised learning when machine learning methods are applied to the fault detection, an (AI) artificial intelligence module is first trained on the historical data which described the technical state of the machine

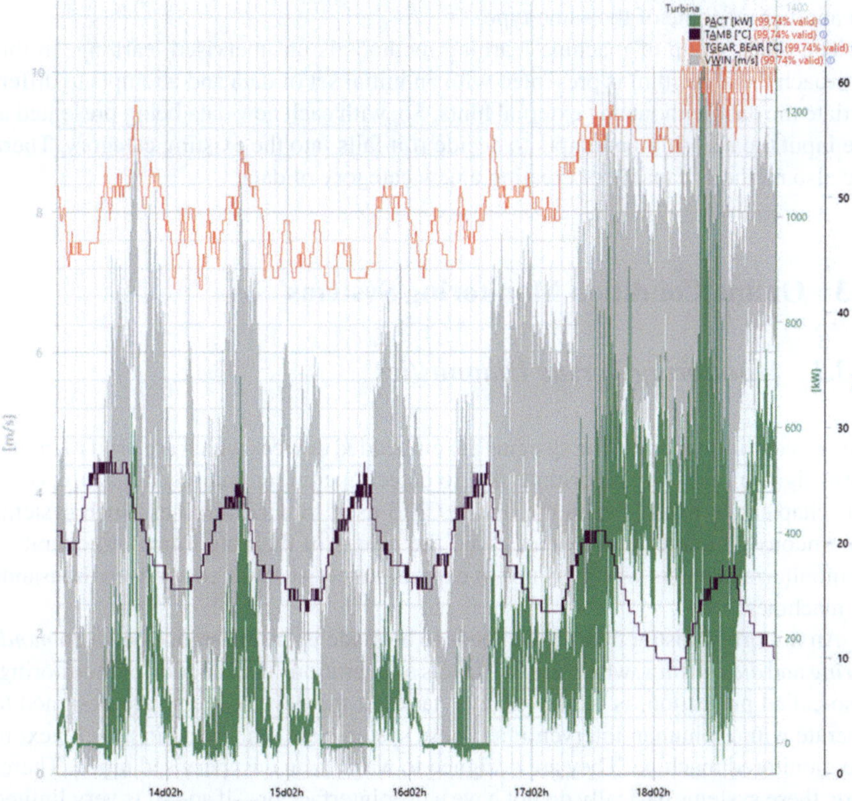

Fig. 3.11 Trend plot showing high resolution history of selected tags from a wind turbine

over a certain period of time. It typically includes lots of "healthy" data. The AI module then uses the data to learn the signal patterns in given technical states. After the training period is completed the algorithm performs the classification task which means it processes the current data sample and finds the match using the known data from the training set.

The availability of the training data is crucial for the learning phase of the machine learning algorithm and for its proper operation afterwards. The training data must fulfill requirements which are hard to meet. As far as healthy data is concerned, it must cover all the operating conditions (there must be enough data for combinations of wind speed, output power, rotational speed, ambient temperature, ambient pressure etc.). Only this single requirement demands months if not years of data and is hard to meet. Next, if the AI is supposed to detect a specific fault, an extensive set of training data taken during the faulty state should be provided and it should also cover all the operating conditions. When we try to cover all the possible faults (in several severity stages) and all the possible operating points, one can understand why it is impossible to collect such a dataset. To make things more challenging, the algorithm should be able to generalize results. Generalization means that the algorithm trained on a dataset from a single machine (or a group of machines) will perform satisfactorily on another machine of the same type.

The other group of machine learning methods is unsupervised learning. In this approach the algorithm is presented with an initial set of data and attempts to differentiate the data without any external hints. So, with each new data being presented at the input the algorithm attempts to decide if it falls into the existing category. There are also methods capable of creating a new category of data.

3.3 Online Condition Monitoring Systems

3.3.1 Monitoring Versus Diagnostics

Online condition monitoring systems are embedded devices which are used to implement signal processing algorithms and to detect upcoming failures in machinery. In this chapter we will focus on the online CMS used in wind turbines. Such systems continuously monitor turbines and send the results of their algorithms to operators or monitoring centers where groups of experts supervise hundreds or even thousands of machines.

An important distinction which needs to be made is the difference between *monitoring* and *diagnostics*, which are two levels of operation. The function of monitoring, also called protection, is the lower level task. At this level systems are designed to operate without human intervention. These systems are embedded in a rack next to the monitored machine. They are designed to work in an unsupervised mode. Therefore, these systems basically do not have a user interface or—if so—it is very limited and used only occasionally to check whether the system is functioning properly and whether there are any active alarms. Since the reliability is of a primary concern, these systems tend to use simpler signal features. The primary goal is safety of the machine and not necessarily detection of a fault at its early stage. If a machine fault is detected, the monitoring system is supposed to autonomously evaluate the signal, generate the signal features and detect the threshold violation. The detection is signaled with binary switches or with a digital interface to a turbine controller. The controller then can decide about automatic action even to the point of stopping the machine.

The diagnostics is the higher level task. It is not so much focused on protection, but rather on the early detection of faults. In the balance between false alarms and early detection, we move more towards the early fault detection. The diagnostic systems involve calculating numerous advanced signal features, e.g. narrowband rolling bearing features. The system tracks trends of features and is able to detect early signs of technical state deterioration, even when the machine is still perfectly functional. Instead of automatic connection to the controller which can switch off the machine, the diagnostic systems send notifications to higher level systems. It is perfectly acceptable as faults are supposed to occur at an early stage and so, an e-mail report is a perfect means of communication.

This distinction between monitoring and diagnostics was originally proposed for other industries. It is common in large power generation and in oil and gas industries. For example, the API670 [11] standard explicitly requires that these two layers should be separated into different computer systems and the failure of the diagnostics layer must not compromise the operation of the protection layer. Such an approach increases the reliability and availability of the machinery, but at the cost of a more expensive condition monitoring. In the wind power generation industry it is common to mix these two functions in a single system. The distinction is provided on the level of an embedded system installed next to the control system in the nacelle and on the level of the vibration expert software which is run remotely by a human expert.

3.3.2 General Structure of CMS

The general structure of the Condition Monitoring System is presented in the Fig. 3.12. Every CMS has a measurement module which first performs signal conditioning. Signal conditioning is the adjustment of physical electric signals (e.g. IEPE standard used for piezo accelerometers) to the electric standard of the Analog Digital Converter (ADC). Additionally, it may provide a protection circuitry, so called optoisolation. With this function, input signals are galvanically isolated from the rest of the system and in case of an electrical surge on an input, the main part of the CMS will not be affected or even damaged. After signal conditioning signals are sampled with the ADC. At the output we obtain sampled vectors of raw vibration data.

Raw vibration data cannot be directly used to assess the technical state of a turbine. The main reason is too large an amount of data. Each sampling session of a few

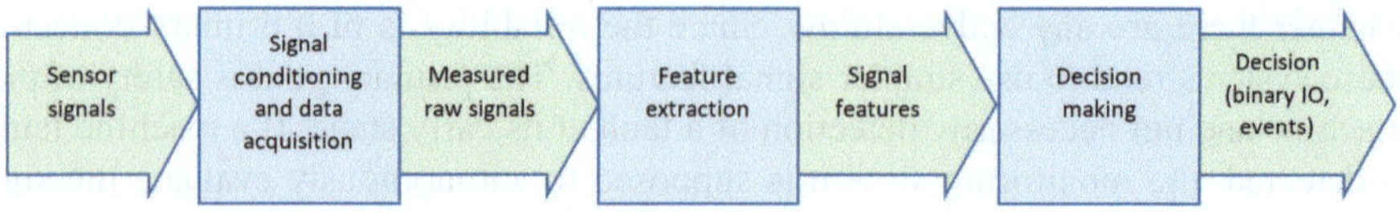

Fig. 3.12 General structure of condition monitoring system

seconds at the sampling frequency of dozens of kilohertz will produce hundreds of thousands or even millions of raw samples per a single vibration channel. There are even systems which sample signals for hundreds of seconds as required for a slow rotational speed machinery, like planetary gearboxes. Since CMSs are multi-channel systems, the amount of processed data is very high and requires efficient and powerful processors and large memory circuits. To make decision making possible, raw vibration data needs to be processed. This is achieved with algorithms, as discussed in the Chap. 2. Each algorithm uses raw vibration data as its input and produces a single number at the output. All the algorithms are grouped in two sets, namely broadband features (e.g. rms, zero-peak, kurtosis) and narrowband features (e.g. energy in a defined frequency band or order band). Some of these algorithms require extensive processing and calculation of Fast Fourier Transform, signal resampling or signal envelope. All these requirements bring additional demands for computing power and available memory.

After the feature extraction step the data obtains the form of a table where each row contains all the calculated features and columns are the consecutive moments in time. Such data format allows for decision making algorithms to run. The result of this step is a set of decisions either in the form of a binary output or an event stored or sent to an external system like SCADA. The simplest form of decision making is thresholding. The value of the feature is checked against the alert and alarm thresholds and if a violation is detected it is stored in a database and an operator is notified. It is also possible that a digital output is activated in such a case.

3.3.3 CMS Architecture

Figure 3.13 presents the architecture of CMS installed on two wind farms. Each CMS is mounted in the turbine nacelle, next to the drivetrain. All the sensor cables are connected directly to the CMS. Typically, such a system has dynamic inputs for vibration signals, process inputs for wind speed and generator power and the tacho input for the rotational speed sensor. Often, there is a CMS server on the farm to collect the history of operation. The server can also provide a user interface to the system.

Data from the wind farm is transferred to the control center which usually has such a location that it can serve several wind farms. The user interface in the control center is focused on operators who need up-to-date information about the current

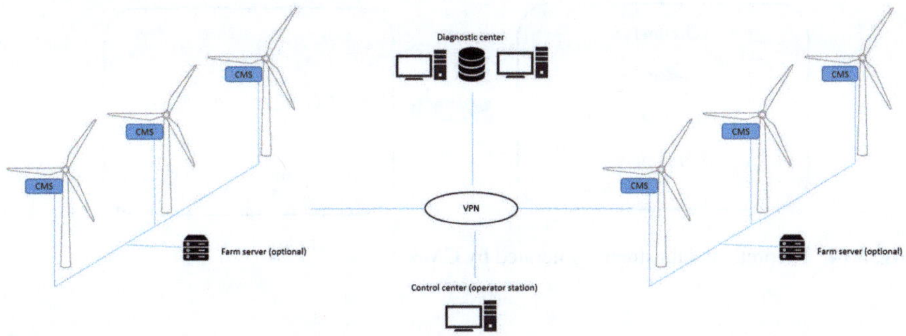

Fig. 3.13 Architecture of CMS installed on wind farms

health state of their wind turbines. Finally, the data is transferred to a diagnostic center. Diagnostic centers employ a group of specialists who are expert at advanced machinery monitoring services. The amount of collected data allows to run algorithms comparing individual machines across the entire fleet of the same type. It is then easier to quickly detect outliers that should be taken care of without delay. The databases collected in diagnostic centers are a perfect environment for developing new health assessment algorithms. From an organizational point of view, the center may be operated by three different groups of organizations: turbine owners, turbine manufacturers or CMS manufacturers. This certainly has a large impact on the set of goals of the particular center.

The Condition Monitoring System has the following functions:

- signal conditioning,
- data acquisition from the sensors,
- signal validation,
- calculation of signal features,
- detection of alarm threshold violations,
- storage of the data in the database,
- operator notification about detected malfunctions.

Storage of measured data needs to be given a special consideration. On one hand, collected data is the most valuable asset of the CMS and forms the basis for decision making and more advanced data analysis. On the other hand, storage of raw vibration data is not feasible due to large volume of data. To better understand the data stream refer to Fig. 3.14.

The data rate of 400 KB/s equals the stream of 4 Mb/s and is in the range of a HD movie. Moreover, continuous storage of such a stream would require memory space of 35 GB/24 h and 12.5 TB per year. Still, the numbers refer to a single wind turbine only and the CMS data centers store data from hundreds of individual CMS systems.

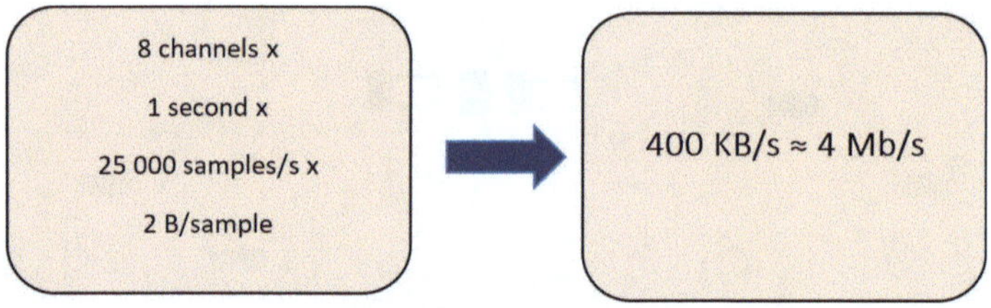

Fig. 3.14 Theoretical data stream generated by CMS

Luckily, there is little benefit in storing all the raw vibration data. The dynamic behavior of the drivetrain components does not vary much, except in a faulty condition. In practical CMS solutions, it is sufficient to store the values of signal features only, for example every 10 min (this is a configuration setting). With this approach, assuming 150 features per CMS, the amount of stored data goes down to 43.2 KB per day which equals 15.6 MB per year. This means 6 orders of magnitude reduction in the data storage requirements. Having said this, one has to admit that there is a very large value in the raw vibration data after all. If an exhaustive vibration analysis is to be done by an expert, the raw data are the most complete source of information (together with the process data which provide the context). A typical solution to this problem is to store a record of raw data only when something extraordinary happens, like the threshold violation. Apart from that some systems offer to store a raw data record every day or every week.

3.3.4 CMS Operator Stations

Operator stations are desktop PCs which present the data to the user. There are various IT technologies at hand which are used for this purpose. The traditional approach was to develop an application running under an operating system, e.g. Windows. It offers large functionality, but requires installation of the software, receiving and installing updates. Another very popular technology is the user interface running inside a web browser. Such an approach has some limitations in its functionality, but offers a very easy access to the application and its updates. With the rapid ascent of mobile devices there are also user interface versions for tablets and smartphones.

There are two distinct groups of users, namely operators and vibration experts. The operators need to quickly access the data describing the current status of the machine. Thus, the typical screens are a mimic screen (see Fig. 3.15), an event list, a data table, a trend plot.

Vibration experts require access to signal analysis tools which are plots, for example: frequency spectrum, order spectrum, signal envelope, XY plot etc. The screen is often expected to present several pieces of information at the same time. Figure 3.16

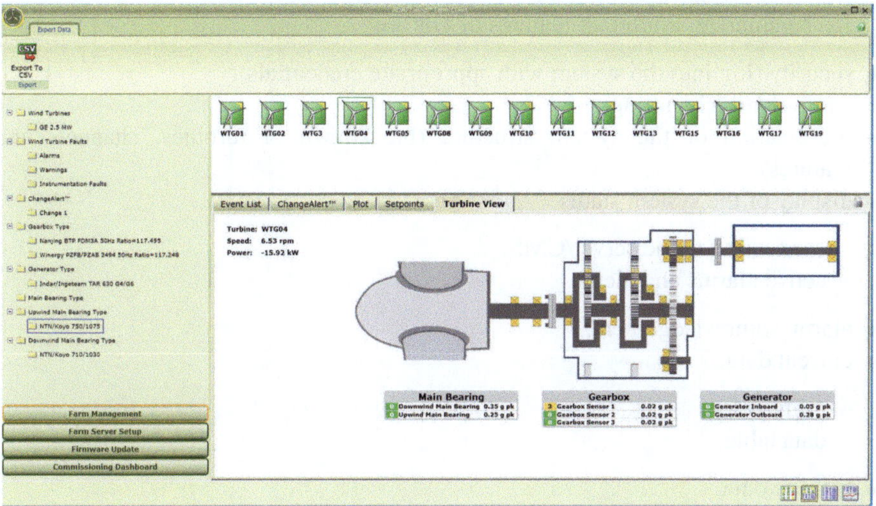

Fig. 3.15 Mimic plot of the commercial CMS

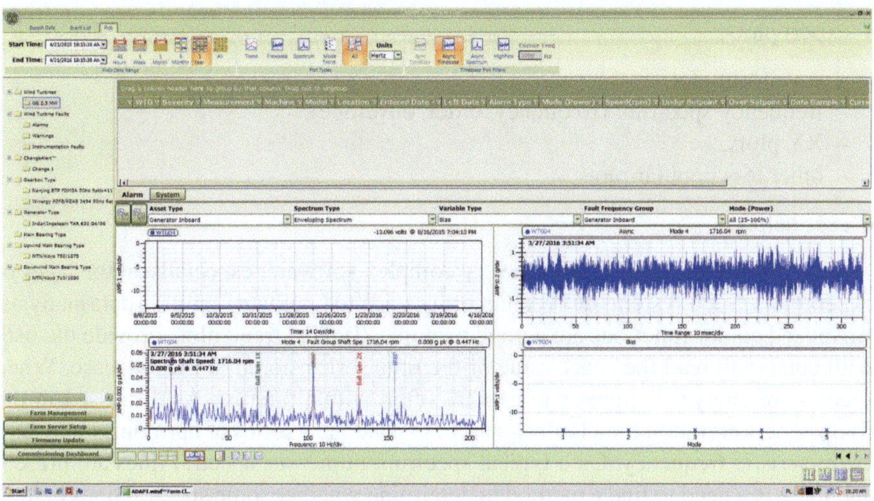

Fig. 3.16 User interface of the commercial CMS with a set of expert tools

depicts the example of a screenshot with a set of expert tools. On the four panels there are: a trend plot, a waveform plot with a raw vibration signal, a spectrum plot and a bin plot which presents distribution of a selected signal feature for 5 power bins.

To summarize, important features of UIs are:

- security (log into the system with appropriate credentials),
- configuration handling,
- presentation of the system structure (breakdown of turbines, channels and features),
- display of the system status,
  - connection to the server/CMS,
  - active alarms and alerts,
- alarm acknowledgement,
- current data,
  - mimic plot,
  - data table,
- historic data,
  - event list (alerts and alarms),
  - trend plots,
- expert plots,
  - waveform plot,
  - frequency spectrum (frequency, order, envelope),
  - XY plots,
  - other less popular plots,
- user help.

User interfaces of the CMS are very complex software, especially in the layer of expert plots. Such systems deliver numerous tools to facilitate vibration data analysis. Advanced cursor functions are a good example. Typical trend plots provide the user with cursors to read the exact value of the time series stored in the database. When spectra are analyzed, signal patterns are often within interest. As discussed in the Chap. 2, it is common for a fault signature to generate a series of harmonics of a fault characteristic frequency. On a typical spectrum numerous spectral lines are present and there is a need to find a method to determine whether some spectral lines belong to the same family of harmonics. The easy and popular tool for this purpose is the harmonic cursor. It is a set of cursors which are spaced by the value of harmonic frequency and are linked together, so moving any cursor with a mouse causes the whole family to move maintaining the equal spacing between them. The example of its usage is presented in Fig. 3.17.

In this case, the cursor was used on the order spectrum, i.e. on the signal after resampling. A family of very sharp spectral lines is visible with strong three first harmonics. The fourth one is much smaller, but still visible. Typical functionality of the harmonic cursor includes adding and deleting additional lines. Also, exact values are visible for each cursor line. It should also be possible to show the exact value of the spectral line amplitude.

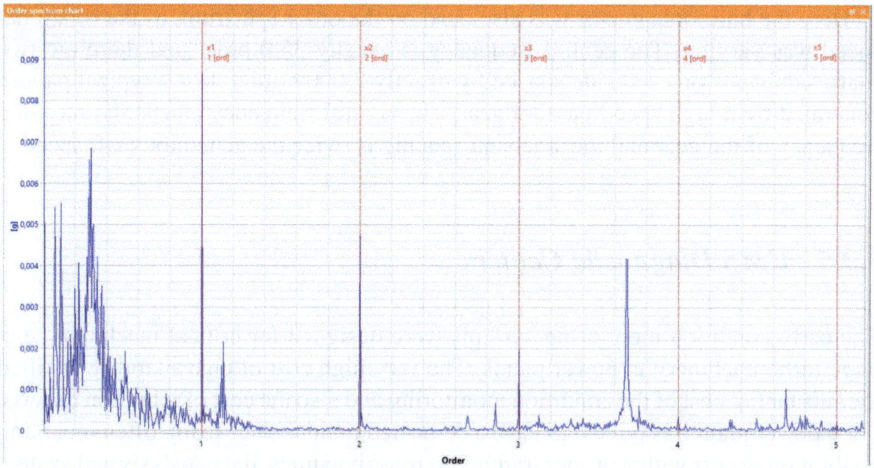

Fig. 3.17 Example of a harmonic cursor

Another similar tool is the sideband cursor used for analysis of the gears. Since faulty gears generate sidebands on the spectrum there is a need to detect which spectral lines belong to one sideband family. This cursor type has an additional degree of freedom compared to the harmonic cursor. Apart from the harmonics, the sideband cursors can be placed on the spectrum. By moving a selected line, all the other lines of the family move accordingly. The example of the sideband cursor is presented in Fig. 3.18.

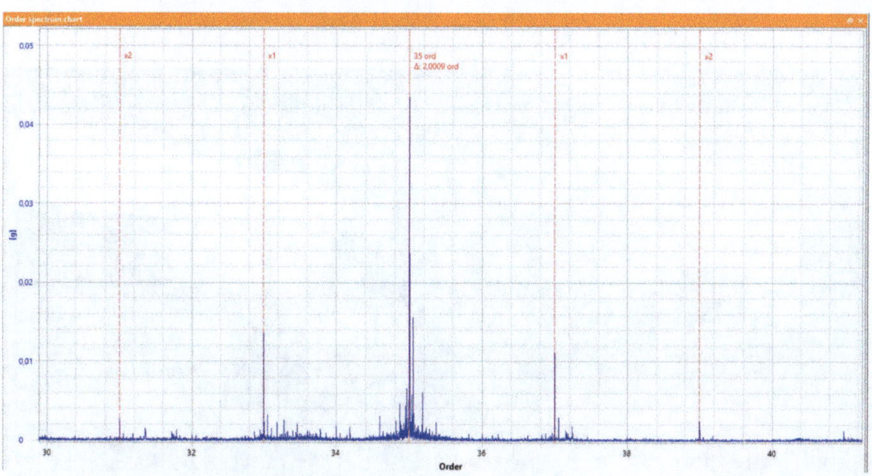

Fig. 3.18 Example of a sideband cursor

Here the harmonic cursor was also used on the order spectrum, as the rotational speed was varying. The central frequency is exactly 35.0 order and there are two sharp sidebands on each side spaced by exactly 2.0009 order. It is a typical pattern generated by a high speed parallel gear. The central frequency equals the meshing frequency of the gear and the sideband spacing is twice the generator shaft speed.

3.3.5 CMS Diagnostic Centers

The early condition monitoring systems were designed for critical machinery, e.g. large turbomachinery in power plants. The very high cost of unavailability justified the substantial cost of the condition monitoring and also the cost of vibration experts. For a power plant it is typical to have a dedicated maintenance team, often including a vibration expert with a proper training in rotordynamics, data analysis and system usage. With the number of CMS installed on wind turbines growing to hundreds and thousands, this model turned out to be no longer economical and a single expert was needed to oversee higher and higher number of turbines. With advances in telecommunication and lowering costs of data transfer, the answer to this problem was foundation of diagnostic centers. The example of such a center is presented in Fig. 3.19.

Diagnostic centers collect data from CMS servers on wind farms or on individual turbines. To further limit the involvement of people, several automated tools are used to scan data. The attention of an operator should be drawn only as seldom as necessary.

Apart from that, the user interface is similar to that one for an expert on the wind farm level. The default information is the event list and map which helps to pinpoint

Fig. 3.19 Example of a diagnostic center (with permission from BaxEnergy GmbH)

any abnormal behavior. When needed, one can quickly run the analysis tools. Many providers of the systems try to develop tools which will be very easy in use and the expected task should be achieved in the least possible number of clicks. The other important functionality is the reporting. As the technical state and analysis steps should be properly documented, such software often provides the possibility to automatically generate the periodical and alarm reports.

3.3.6 CMS Configuration

Probably the most important consideration about the CMS is its configuration. Actually, proper configuration of the CMS is more important than its actual type, as most CMSs share a large part of functionalities. According to Dvorak [12], the key issues when using CMS are:

- missed detection of failures,
- false alarming or too many alarms,
- spending several months up to a year or more on adjusting alarms,
- not being able to detect planetary gearbox issues,
- not enough lead time prior to failure.

These issues are often caused by lack of a proper system configuration. That is why, it is so important to understand the configuration in detail.

Early CMS systems (as well as other computer systems) were configured by editing text files. Currently, the configuration information is well structured and stored in an SQL database or a set of XML files. The information is accessed by a specialized configuration module. Such a module can be a separate application or a part of the user interface application depending on the developer's choice. To be able to edit the configuration, one should have sufficient access privileges. Good quality CMS provides a module for user management with appropriate user rights. A configurator module provides following functions:

- creating and modification of system users,
- adding new hardware units,
- editing hardware settings (sampling frequency, sensitivity, sensor type etc.),
- definition and modification of calculated features (frequency bands, averaging, etc.),
- definition and modification of threshold levels,
- definition of mimic screens,
- definition of pre-defined screen sets,
- check of configuration consistency.

The goal of the configuration process is to match selection and settings of features to the monitored drivetrain. Only the correct configuration allows for correct operation of the CMS, and any errors made at this stage will compromise the overall machinery maintenance process in the company. To facilitate the configuration and reduce the number of faults many systems include software tools to help in the configuration process, especially in defining features and setting the alarm threshold levels.

A specific feature of geared wind turbine drivetrains is their complex kinematic chain. It consists of several shafts connected with toothed gears. The rotational speeds of shafts differ up to the factor of 100, from a dozen rpm for the main bearing to 1800 for the generator shaft. Additionally, all the shafts have rolling element bearings which add specific vibration signatures with high frequency content to vibration signals. A rotational speed sensor is typically installed on one shaft only—the generator shaft, which has the highest speed and it needs to be recalculated for all the other shafts in the turbine.

The configuration process needs to be automated and facilitated by software in order to be efficiently used by the user. The main trouble is the effort or number of man-hours it requires, as even in simpler wind turbines there may be about 150 signal features and several hundred thresholds to set up. A very important task in the configuration is defining all the drivetrain characteristic frequencies (see Chap. 2). Based on this information the software should automatically calculate the band frequencies of narrowband features. A specialized tool which is capable of creating the drivetrain from blocks (shaft, REB, parallel gear, planetary gear, rotor, etc.) is of a great help. Figure 3.20 presents the example of a software tool to graphically build the drivetrain.

The other part of the configuration process is the threshold value setting. As described in the Chap. 2, the current norms only propose values for the velocity rms of onshore wind turbines. All the thresholds for configured signal features need to be set based on the experience. The primary source of this knowledge is the experience taken from other machines of the same type or (if it is unavailable or in order to get more detailed values) the operational history of the turbine. Due to highly varying operating conditions the best approach is to create so called "operational states" (or

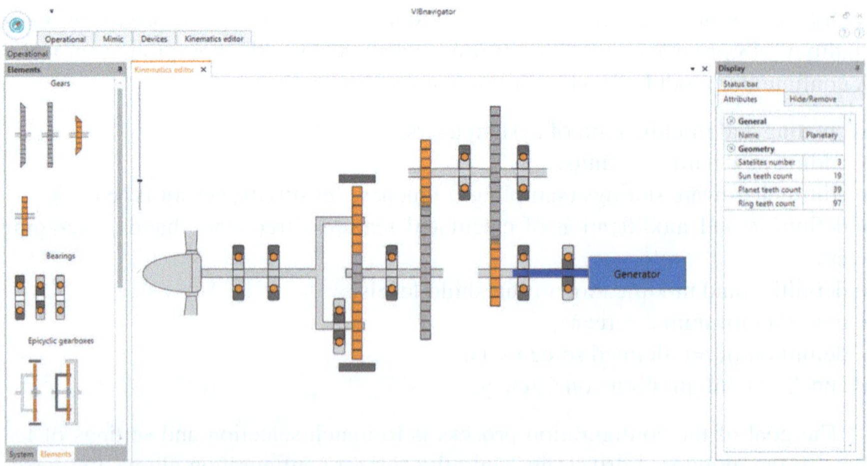

Fig. 3.20 Example of a software tool to graphically build the drivetrain

bins), analyze only data falling into these bins and then calculate the values. The discussion on mathematical algorithms used to set the threshold values can be found in Jablonski et al. [13]. As shown by Jablonski et al. [14] the automatic validation process depends on "cleanliness" of the collected data. Since the data may be affected by numerous sources of noise, one should always validate the data before attempting to set an automated process of threshold.

3.3.7 CMS Examples

There are many condition monitoring systems available on the market. The division of CMSs into several market segments is an interesting task on its own, but it is not the purpose of this book. It is only worth mentioning that monitoring systems range from simple single channel devices starting at a few hundred euros to very sophisticated protection and diagnostic systems for power plant turbogenerators with a price tag of hundreds of thousands of euros. Since this book is focused on wind turbines only examples of relevant systems will be presented.

A typical CMS is installed in a turbine nacelle, in a control system rack or in its own (next to the turbine controller). Figure 3.21 presents such a typical condition monitoring system suited for wind turbines.

The CMS is installed in a cabinet separate from other systems. The protective enclosure holds the safety fuse connection (bottom left), the power supply (top left), the connection bar for cabling from sensors (right bottom) and the acquisition unit (right middle). The CMS may be connected to the turbine controller to get the process data, to the SCADA system to provide it with the technical state features and to a higher level diagnostic center.

The manufacturers of the systems strive to keep their products up to date and we can witness many new products on this market. It is vital that the reader will have the understanding of the already large and mature market. It is steadily growing though and the market survey shows that it will experience a stable growth over at least the next 5 years, as shown by MarketsandMarkets [15]. There are teams collecting information about CMSs available on the market and the interested reader can refer to the report by Crabtree et al. [16].

The selection of the systems presented in this book is not exhaustive because of too big a number of them. Therefore, the examples presented on the following pages are simply my subjective choice. The choice was not based on a sophisticated survey, but it is rather a short list driven from my personal experience in this field, supported by the manufacturers' web resources. To show the CMS world from different angles I tried to present one example from the following groups: companies specializing in vibration monitoring, large wind turbine manufacturers, control system manufacturers and finally an interesting innovative approach using MEMS sensors.

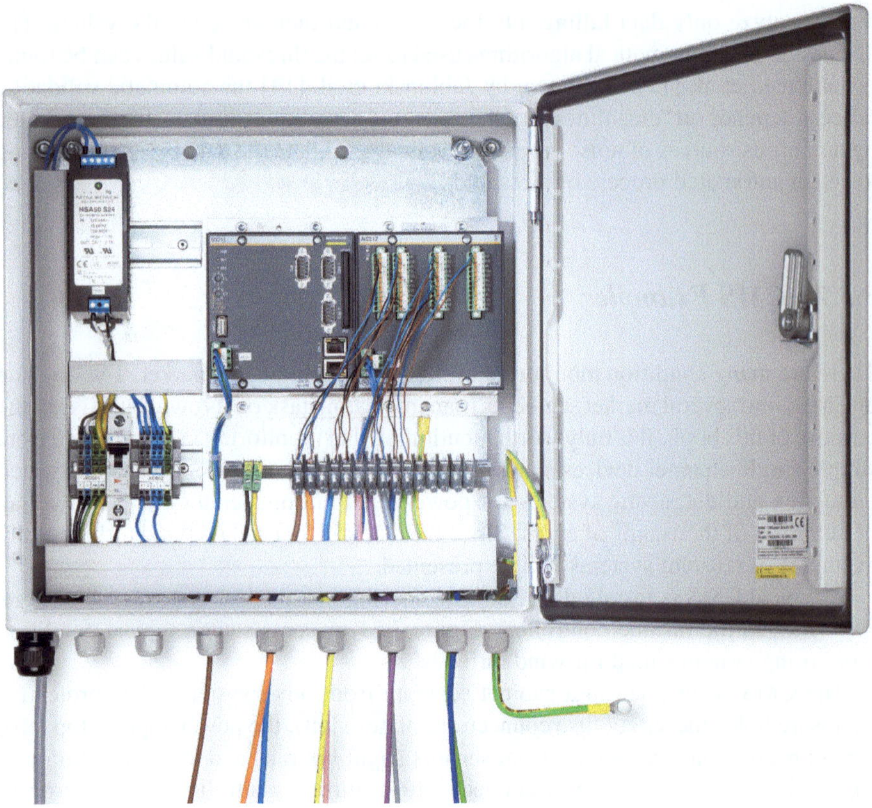

Fig. 3.21 Example of a condition monitoring system: Ω-Guard® system in standalone configuration (with permission from Bachmann Electronic)

The first example is the *WTAS* (Wind Turbine Analysis System) system from *BK Vibro*. The company is based in Denmark and takes advantage of many decades of experience in vibration monitoring under the name of Bruel & Kjaer. The WTAS system consists of a unit installed in the nacelle and the accompanying software. Key technical parameters of the system are presented in Table 3.4.

The WTAS system is a unit suited for a vast majority of wind turbines on the market. It is designed to collect the data onsite and send it to a center for detailed analysis. The data acquisition module has high parameters. It features 24 bit Analog Digital Converters and is able to perform simultaneous acquisition on all the channels. The configuration of the system can be edited in a graphical way, where signal processing steps are depicted as building blocks which need to be connected to create a so called monitoring template. WTAS is capable of acquiring and storing raw vibration waveforms from up to 16 channels. Storage can be performed on a hardware trigger (digital input), software trigger (detected alarm) or user request. Additionally, raw buffers can be stored as pre-trigger (given amount of time before

Table 3.4 Key technical parameters of WTAS Wind Turbine Analysis System

Parameter	Value
Number of analog inputs (AI)	16 differential
Sensor types compatible with AI	Vibration, tacho, temperature
Dynamic range	90 dB
Resolution	24 b
Frequency range	0.1 Hz–10 kHz
Sampling frequency	25.6 kHz
Digital inputs	4 galvanically isolated
Digital outputs	2 SPDT relays 24 V, 100 mA
Network interfaces	4 Ethernet (3 RJ45, 1 optical)
Serial interfaces	1 RS-232, 1 RS-485
Environmental	IP66
Operating temperature	−40 to +55 °C

the trigger), post trigger (given amount of time after the trigger) or a combination of both. This helps the vibration expert to analyze in detail the root cause of an event. Calculated signal features (called characteristic scalar values) are stored in the WTAS buffers and can be accessed by the web based software WTG Analyzer. The data from WTAS is periodically transferred to the central data acquisition server at the Remote Monitoring Center. The manufacturer states that 7000 turbines are being monitored in the BKV center this way.

Another example is the *ADAPT.wind™* system from *General Electric*. The company is based in the USA (although its numerous other branches are abroad) and is one of the world's largest manufacturers of wind turbines. The ADAPT.wind system is the in-house development based on over 50 years of experience in vibration protection and monitoring. This expertise was greatly advanced when in 2002 GE acquired the Bently Nevada company, the pioneer and leader in vibration based condition monitoring. The ADAPT.wind system consists of a unit installed in the nacelle and the accompanying ADAPT.wind web based software. Key technical parameters of the system are presented in Table 3.5.

ADAPT.wind can be used on GE wind turbines and is also applicable to other turbines. The data acquisition unit has very good features, with high dynamic inputs (110 dB) and sampling frequencies of up to 102.4 kHz. DAQ module can be interfaced to the SCADA system, e.g. to get the generator power value or to return turbine health status. The system can be operated in three modes: as an independent standalone system, as a data acquisition part of a networked system or as part of GE's OEM package. The second and third options reveal the full functionality of the system. The user can then access the data with the web based ADAPT.wind software. The configuration includes a large number of preconfigured turbines, so the user does not need to select his own. The data is analyzed depending on the generator output power which divides the data into 5 bins reflecting 20% ranges from 0 to the nominal output

Table 3.5 Key technical parameters of 60M100 data acquisition unit used in ADAPT.wind™ condition monitoring system

Parameter	Value
Number of analog inputs (AI)	12 dynamic
Number of tacho inputs	2
Dynamic range	110 dB
Resolution	24 b nominal
Frequency range	0 Hz–40 kHz
Sampling frequency	Up to 102.4 kHz
Network interfaces	2 Ethernet (RJ45, 1 local static, 1 network DHCP)
Operating temperature	−40 to +70 °C

power. As a differentiator, ADAPT.wind offers some additional signal features aimed at detection of planetary gearbox faults (Planetary Cumulative Impulse Detection, using an optional oil particle sensor) or variable speed gears (Dynamic Energy Index, an algorithm detecting excessive sidebands).

The third example is the *Ω-Guard®* system from *Bachmann*. The company is based in Austria and is one of the world largest manufacturers of control systems for wind turbines. Bachmann systems are popular on the market and are installed on many wind turbines around the world. This example shows yet another route to approach the condition monitoring market. Bachmann core business is control system and they have decided to extend their offering with the CMS. Close integration with the control system brings several advantages, e.g. interfacing to the control system to exchange the data is much easier in such a case than with a 3rd party control system. Nevertheless, it is interesting to mention that to accelerate the growth in the CMS field Bachmann decided to acquire μ-Sen GmbH. It was a German manufacturer of CMS and Bachmann created its monitoring business out of it.

The Ω-Guard system is based on over 15 years of experience in vibration monitoring of wind turbines. Similar to the other presented systems, Ω-Guard consists of a unit installed in the nacelle and the accompanying WebLog web based software. Key technical parameters of the system are presented in Table 3.6.

Table 3.6 Key technical parameters of Ω-Guard® system

Parameter	Value
Number of analog inputs (AI)	12 (9 IEPE and 3 ± 10 V)
Number of tacho inputs	2 counter, 1 incremental
Resolution	>17 bit
Sampling frequency	50 kHz
Network interfaces	Ethernet, FASTBUS
Operating temperature	−25 to +55 °C

Ω-Guard system can be used in three versions: standalone, Top-Box integrated and fully integrated. In the first one, the system is a complete cabinet with a functioning CMS. Such a solution can be used where there is no Bachmann controller installed, for example as a retrofit of the existing WT. The latter two versions are integrated with the wind turbine Bachmann controller. The data acquisition unit has good technical parameters. Differentiating innovation is the "μ-bridge" sensor developed in house and used for monitoring of structural bending. It is useful to monitor the main bearing and the planetary gearbox where due to very low rotational speed the vibration amplitude at low frequencies is very small and the casing deflection signal has a better signal-to-noise ratio. The data acquired and analyzed by the Ω-Guard unit is then sent to the WebLog Expert server where it is analyzed by experts and can also be accessed by authorized users. The system accommodates all the common signal processing methods and provides quick information about a detected threshold violation. The WebLog has wide set of functions for fleet analysis where the data from many wind turbines can be compared and any outliers are found more easily. An interesting option is the ability to acoustically listen to the vibration signal which may help an experienced vibration analyst to pinpoint the cause of the problem. According to the manufacturer's information, 3500 Ω-Guard systems are monitored remotely from their office in Rudolstadt, Germany.

The fourth and last example is the *ecoCMS* system from *Romax*. The company is based in the United Kingdom and for many years acquired experience in the field of modelling, analysis and design of rotating machinery. The ecoCMS system is probably the newest development out of the presented four and uses another sensing technology, namely the MEMS vibration sensors. Like the other presented systems, the ecoCMS consists of a unit installed in the nacelle and the accompanying software. Key technical parameters of the system are presented in Table 3.7.

The ecoCMS has a different underlying technology which can be shown in the system description and the published data. Since MEMS sensors are less costly than traditional accelerometers, the economy approach and low cost are stressed by the manufacturer. On the other hand, MEMS sensors have a higher noise level and therefore it was decided to use ADC with less resolution than in other systems. Also, the number of other inputs was limited to avoid additional costs. The manufacturer

Table 3.7 Key technical parameters of ecoCMS condition monitoring system

Parameter	Value
Number of analog inputs (AI)	9 non-synchronous
Sensor types compatible with AI	MEMS
Tacho input	1
Resolution	16 bit
Network interfaces	Ethernet RJ45
Environmental	IP66
Operating temperature	0 to +60 °C

underlines the modern architecture of the system and informs that it embraces the Internet of Things approach. The system is affordable, there are minimal upfront costs and there is the cloud based software platform in the center. The software platform was named Fleet Monitor™ and it constitutes the core of the system. The software consists of a central server which acquires and processes the data from ecoCMS units and of a web client. The latter part offers a rich set of plots for operators and vibration experts. The company also offers remote condition monitoring services and informs that it monitors 40% of UK's offshore fleet.

Once again, the author would like to underline that the purpose of the presented selection of CMS systems is purely illustrative. He has no intention of convincing any potential reader about advantages of any of the presented systems over the other ones (or to show advantages over other numerous systems which are not presented here). The main purpose is to present different types of CMS solutions offered by different companies with different backgrounds.

3.4 Portable Vibration Analyzers

3.4.1 Differences Between Off-Line and On-Line Systems

Introduction of an online condition monitoring program on all wind turbines requires a considerable investment in both hardware, software and labor. Often, wind turbine users would like to start the condition monitoring journey without having to bear high costs from the very first day. A portable vibration analyser may be a solution to this problem.

Portable vibration analyzers are a separate branch of the condition monitoring evolution tree. They share much of the functionality but also differ significantly in several aspects. Let us first look at similarities. Portable analyzers are basically condition monitoring systems which can be installed temporarily (from hours to weeks) on a turbine. In such a case, they should be easy to operate and have a functional configuration system. In addition, they should facilitate easy change of the configuration and the ability to come back to one of previous ones. The accompanying software provides all the typical plots of a CMS and is fully equipped in tools for the vibration expert, but it is not going to be used by operators. The acquired data is transferred to the PC-based software. If there are manufacturers who have both portable and on-line systems in the portfolio, they will most often share the same software platform to store and analyze the data.

The differences between portable vibration analyzers and stationary CMSs can be easily spotted. The portable analyzers are used by vibration experts and not operators. They are often installed on a machine which is known or suspected to have some problems and thus it is a vibration expert who will be selected for the investigation job. The configuration of the analyzers is more flexible, as it should enable quick reload of the configuration. All the stored data should remain intact and be available for further analysis and comparisons. Another important feature of the portable analyzers is limited channel count, often being fixed at 2 plus the rotational speed input. To be able to measure large machines the analyzer has an ability to configure the measurement track. The track is a list of all the vibration measurement points on the machine. The technician will then attach each sensor (or a pair of sensors) for the time required to collect vibration data and then proceed to the other location and so on until all the configured points are measured.

The key difference is that a portable analyzer will not provide full protection for the monitored turbine. With one analyzer per the whole wind park we can only monitor one turbine at a time. If anything unexpected happens—and it sometimes does—or the fault is intermittent and is not repeatable, then bringing the portable analyzer may not result in collection of meaningful data. Moreover, if the analyzer has only two vibration channels it takes 3 sessions to cover a typical wind turbine setup with 6 vibration sensors. Since the operating conditions vary greatly the collected data cannot be compared easily but there will always be a difference caused just by different wind speed or a wind gust which happened to occur.

As far as economic factors are concerned, the portable analyzer approach requires smaller initial expenditures (though higher than a single CMS), but in the long run it will mean more labor cost. Human labor is required to reconnect the sensors and the unit, bring it down and up again to another wind turbine and so on, as long as the portable analyzer is used.

3.4.2 Portable Analyzers Examples

It is not the purpose of this book to give an exhaustive guide to this part of the market. Therefore, to present the reader with a general concept of capabilities of portable analyzers only two examples of such systems will be described. They base on different approaches. The first example is the portable version of the Bachmann CMS presented in the previous chapter. It is *Bachmann Portable Condition Monitoring System* and comes from Bachmann Electronic from the USA. Figure 3.22 presents the general view of the system. The system is a dedicated development featuring 18 vibration channels and 1 channel for both rotational speed and power.

The second example is the system designed strictly as a portable analyzer. It is *Microlog Analyzer AX* from the *SKF* company based in Sweden. SKF was at its origin the manufacturer of rolling element bearings and later expanded into the field of vibration based condition monitoring. The presented system is a good example of multifunctional high level measurement device. Table 3.8 presents features of the Microlog Analyzer AX.

The Micrpublished Analyzer AX is a very powerfull device suited for tasks outreaching wind turbine diagnostics only. It is certified to work in hazardous environments when there is a threat of explosive atmosphere. It can accommodate very wide range of sensors including dynamic pressure making it useful in e.g. reciprocating compressor or ignition engine diagnostics. The user can select many options for data processing.

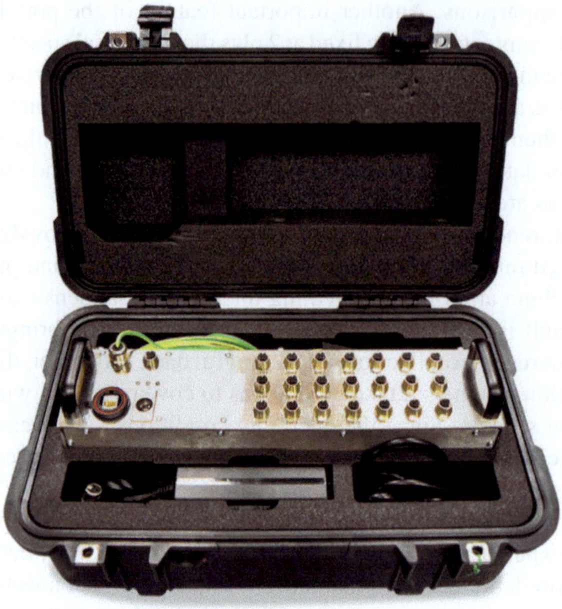

Fig. 3.22 Bachmann Portable Condition Monitoring System (with permission from Bachmann Electronic)

Table 3.8 Key technical parameters of Microlog Analyzer AX

Parameter	Value
Number of analog inputs (AI)	4 dynamic
Sensor types compatible with AI	Vibration, DC, dynamic pressure, temperature, tacho, keyboard entry
Dynamic range	>90 dB
Resolution	24 b
Frequency range	DC–80 kHz
Sampling frequency	25.6 kHz
Environmental	IP65
Operating temperature	−10 to +50 °C
Battery life	8 h
Weight	1.6 kg

There are many options for signal averaging and signal windowing. Data acquisition may be triggered by an external trigger with a selectable slope and amplitude. There are also advanced visualization options available with plots like dual channel spectrum, phase table, peak hold averaging and others. The differentiator for Microlog is SKF Acceleration Enveloping technology (gE) which is described as a very efficient method of rolling element bearing fault detection. Apart from vibration analysis, the device also offers one- and two-plane balancing, which is the function required by many field service and commissioning experts. Microlog can bring a great value to its user provided they are educated well enough to be able to grasp this value.

The data collected by Microlog Analyzer AX is transferred to the SKF software package. The name of the package is @ptitude Analyst and it is a common platform for all the SKF portable and online systems for data storage, processing and visualization. It combines functionality for configuration, route planning, operator level data display and a wide range of vibration expert analytic tools.

As for the online systems, the author would like to stress that the purpose of the presented selection of portable analyzers is of a purely illustrative purpose. He has no intention to convince any potential reader about advantages of one of these systems over another one (including many systems which are not presented here).

References

1. International Standard Organization (2015) ISO 10816-21:2015. Mechanical vibration—evaluation of machine vibration by measurements on nonrotating parts—part 21: horizontal axis wind turbines with gearbox
2. PCB Piezotronics (2018) Specification of 601A02 accelerometer. https://www.pcb.com/products.aspx?m=601A02. Accessed 23 June 2017
3. Albarbar A, Teay SH (2017) MEMS accelerometers: testing and practical approach for smart sensing and machinery diagnostics. In: Zhang D, Wei B (eds) Advanced mechatronics and MEMS devices II, microsystems and nanosystems. Springer International Publishing, Switzerland
4. Analog Devices (2010) Specification of ADXL001 accelerometer. http://www.analog.com/media/en/technical-documentation/data-sheets/ADXL001.pdf. Accessed 23 June 2017
5. Spence E, Chong J (2016) MEMS comes of age—review of a new broadband 20 kHz MEMS accelerometer for vibration measurements. In: Abstracts of machinery failure prevention technologies conference, Dayton, Ohio, 24–26 May 2016
6. Analog Devices (2018) Specification of ADXL1001/1002 accelerometers. http://www.analog.com/media/en/technical-documentation/data-sheets/ADXL1001-1002.pdf. Accessed 20 July 2018
7. ifm (2018) Specification of VSA006 accelerometer. Available from https://www.ifm.com/gb/en/product/VSA006. Accessed 20 July 2018
8. Thomas MS, McDonald JD (2015) Power system SCADA and smart grids. CRC Press, USA
9. Clarke G, Reynders D (2004) Practical modern SCADA protocols. Elsevier, Oxford
10. Tautz-Weinert J, Watson SJ (2017) Using SCADA data for wind turbine condition monitoring—a review. IET Renew Power Gen 11(4):382–394
11. Americal Petroleum Institute (2014) API Standard 670. Machinery protection systems
12. Dvorak P (2015) The alarming state of condition monitoring in the wind industry. Wind Power Engineering and Development. https://www.windpowerengineering.com/business-news-projects/featured/the-alarming-state-of-condition-monitoring-in-the-wind-industry. Accessed 5 Sept 2017
13. Jablonski A, Bielecka M, Barszcz T (2013) Modeling of probability distribution functions for automatic threshold calculation in condition monitoring systems. Measurement 46(1):727–738
14. Jablonski A, Bielecka M, Barszcz T (2011) Automatic validation of vibration signals in wind farm distributed monitoring systems. Measurement 44(10):1954–1967

15. MarketsandMarkets Research Private Ltd. (2018) Machine condition monitoring market, by product (vibration monitoring, thermography, ultrasound emission, lubricating oil analysis, corrosion monitoring, and motor current signature analysis), Component, Application, and Geography—Global Forecast to 2024
16. Crabtree CJ, Zappalá D, Tavner PJ (2014) Survey of commercially available condition monitoring systems for wind turbines, Technical Report. Durham University School of Engineering and Computing Sciences and the SUPERGEN Wind Energy Technologies Consortium

Chapter 4
Signal Preprocessing and Validation

4.1 Importance of Signal Validation

Despite the rapid development of the technologies of embedded systems, efficient development and deployment of a Condition Monitoring System to a large fleet of wind turbines (often hundreds or even thousands of machines) is often a hard task [1]. The main difficulty is harsh operating conditions which may affect the reliability of CMS system. Such systems can also experience occasional malfunctions of the CMS itself. There are cases when such problems are repetitive and quite frequent. To a large extent the problems are similar to those which happen in other heavy-duty machinery. In these environments, CMS operation can be faced with: large electrical disturbances, sudden load changes, saturation of sensors, unsupervised cable disconnection and finally, costly maintenance—to name only a few. The inherent non-stationarity of wind turbine operation further excludes a major portion of commercially applied diagnostic features from the start. Another key function of CMS systems is data storage for further usage, e.g. in machine learning applications. Today, such databases are measured in terabytes and petabytes. Validation of the data to be stored is a crucial requirement in order not to waste storage for meaningless, corrupted and finally useless data. The importance of the problem is often not recognized, though there are a few authors who put emphasis on the issue. As stated by Yan and Goebel [2]: "vibration monitoring relies on accurate and reliable sensor readings". Consequently, as noticed by Alag et al. [3]: "where the sensors are not significantly more reliable than the systems being monitored, the indication of an abnormal state may be the result of a sensor failure rather than a system failure". This chapter includes work published previously by the author in Jablonski and Barszcz [4] and Jablonski et al. [5]. The content also reflects advances made in the years after publishing these papers.

The goal of signal validation is making a decision whether the acquired signal can be used for subsequent data analysis. In case of wind turbines two conditions need to be fulfilled: a correctness of data acquisition path and a meaningful turbine

© Springer Nature Switzerland AG 2019

T. Barszcz, *Vibration-Based Condition Monitoring of Wind Turbines*, Applied Condition Monitoring 14, https://doi.org/10.1007/978-3-030-05971-2_4

operational state. The first one is hardware related. It involves all the components of the embedded system, namely: accelerometers, cables, a signal conditioning unit, analog-to-digital converters (ADC), a data processing unit, and a data storage unit. In real installations, even if modern CMS systems are very reliable, when monitoring hundreds of machines, each with a dozen of sensors, such a path is likely to experience a data acquisition failure. The data acquisition unit will often recognize an erroneous signal and mark it as invalid, but such a test is often limited to the sensor check only in serious situations when a sensor is saturated or a cable breaks.

The second condition requires a meaningful operational state. Even for relatively simple pumps or fans it is not useful to analyze and store vibration data when the machine is stopped. For the other ones, like large turbomachinery, transient data is of utmost importance for the analysis of rotor dynamics. For wind turbines data quality is determined by the operational state of a machine. In order to store data, the turbine should operate in a defined range of process parameters and the most important ones are the rotational speed and the output power. Vibration patterns highly depend on these values and changes in vibration signals caused by a varying operational state are often larger than those caused by a fault, especially in its early stage. This condition created an important differentiator from classical CMS solutions designed for a typical rotating machinery. The CMS suited for wind turbines applications should allow to define a few operational states and analyze the data accordingly. This subject was discussed in Chap. 3. One should mention that except process channels these two conditions are independent. If speed or power enters a defined operational zone, the final product of a data acquisition path should be stored, but subsequent analysis will be calculated regardless of potential errors of a vibration signal.

Vibration signal validation may be perceived in various ways. Wind turbines are a group of rotating machinery in which acceleration amplitude is used to characterize the machine technical state. If a fresh engineer were to visually assess a number of vibration signals using the time view figures, they would perform the task with a good percentage of correctness simply with an engineer's intuition. Nonetheless, if the solution is to be implemented in an online CMS deployed in hundreds of machines, the key question is how to implement these intuitive rules into an algorithm. Validation rules need to be coded within a programming language or another scientific tool. The list of available tools is continuously growing, and the most promising branches include (to name a few):

- Mathematical rules,
- Signal processing tools,
- Statistics tools,
- Artificial Neural Network (ANN),
- System modelling.

The first logical step is to start with mathematical rules including necessary signal processing and statistics as these methods are relatively efficient in terms of time consumption, easy to implement and have a physical interpretation. All these factors are very important in case of industrial implementation.

The problem of validation of vibration signals is essential for the on-line monitoring systems, in contrast with off-line data collectors. For off-line analysis, signal processing is usually preceded by a quick visual analysis by a diagnostic engineer. This—to a large extent—eliminates the need for automatic computer signal validation. The importance of automatic validation has grown with the implementation of installations of CMS systems on thousands of machines. There are three reasons which drive the research in this field.

Firstly, all the signal processing methods assume the input signal is correct, i.e. free from data acquisition errors. If input data to an algorithm is incorrect, the result is usually unpredictable. It certainly should not be taken into account for decision making about the state of a machine. It is even more important in practice, as hardware generated invalidity messages are insufficient to detect error data.

Secondly, the automatic validation sets up a transition from a single academic test-rig-signal to a large distributed commercial database. These are very different environments and experience gained in the laboratory does not provide all the skills required in the field. On a test rig, it is much easier to provide (typically experienced) staff for the experiment, so the collected signals undergo immediate visual verification and the analysis is carried out only if the signal is recorded correctly. If it is not the case, corrections can be done quickly. It is also easy to select every element of the data acquisition and processing path. Next, the settings of sensors range are set optimal, since there is time to do a number of trials and adjust the CMS to the test rig. On the other hand, in the field the visual verification of the time signal is very seldom (if at all). There is simply an avalanche of data from thousands of sensors. Raw vibration signals are not stored permanently (except for a very small fraction). Thus, if an invalid signal feature is calculated by the embedded data acquisition unit, it is stored permanently, before the signal error is captured and with no chance of verification with the original raw signal. Finally, sensor settings are set by default for a given type of a turbine, eventually being far from optimal. The differences between these environments are summarized in Table 4.1.

Table 4.1 Comparison of data sets from test rig and from field systems (reprinted from (Jab et al. [5] with permission from Elsevier)

Test rig signals	Field systems signals
Immediate visual verification (the analysis is carried out only if the signal is recorded correctly)	Seldom visual verification of the time signal
Each and every element of the data acquisition path can be precisely selected, since there is only a single system to validate	Temporary storage of a signal; if a major data acquisition fault occurs a number of incorrect data-calculated estimators is stored inevitably, before the signal error is captured
Sensors ranges settings are set optimal, since a number of trials is feasible time-wise	Frequently, sensors settings are set by default, eventually being far from optimal

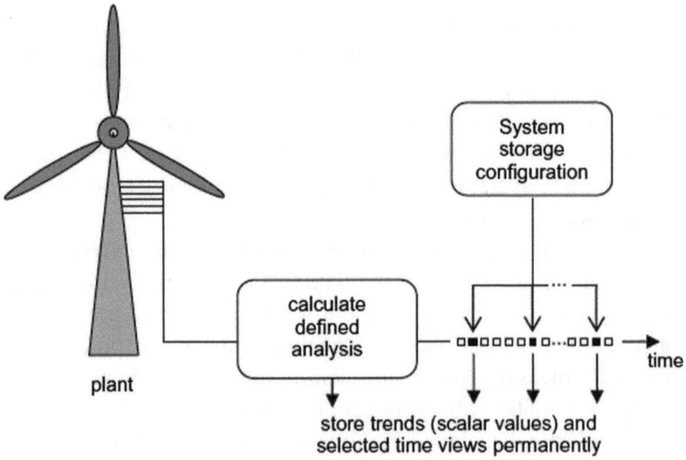

Fig. 4.1 Scheme for selective permanent raw vibration data storage (reprinted from (Jab et al. [5] with permission from Elsevier)

Thirdly, regardless of a set of signal features being calculated, a raw vibration signal is always the basis for any deeper analysis. The basic analyses include time-domain analysis (peak-to-peak, statistical moments), and spectral-domain analysis (FFT, envelope spectrum, order spectrum, narrowband frequency selective features). Since the time data requires a huge data storage space compared with the calculated features, it is typical to store in the database only a few selected signals and delete all the others, right after the calculations are done. Figure 4.1 illustrates the idea of selective data storage.

Here we come to a conclusion on the base of the factors presented. If an incorrect or invalid signal is accepted by the system and the signal features are calculated, the numerical outcome of the signal feature extraction is unpredictable and the chance for data error detection at a later time is very low. The exact chance depends on the ratio of the number of signals permanently stored to the number of all registered signals and it is typically less than 1%. Consequently, the trend value calculated from an invalid feature may ruin the turbine state assessment, e.g. by generating false alarms which is a common problem in condition monitoring systems. The main challenge behind this procedure of data selection is that in case of wind turbines the probability of a rapid change of operational parameters within a single raw record of a long vibration time series is much greater than in case of machines operating under stationary conditions in undemanding arrangement.

4.2 Data Selection and Storage Frameworks

4.2.1 Selection Based on Fixed Time Intervals

As discussed in Chap. 3 the data collected by the CMS is the key resource and basis for assessment of the technical state of a turbine. Collection of all the vibration data requires huge amounts of memory, especially when raw vibration signals should be stored. Therefore, the condition monitoring system continuously calculates signal features and stores them on a predefined basis. The raw vibration data is so large that it is stored occasionally, much more rarely than the features. Before storing the data, enough care should be taken to store only the representative signals. Such signals are usually those stored when the turbine was operating in a defined range of operating conditions and the dynamic behavior was as stable as possible. The layout of signal processing and storage was presented in Fig. 4.1.

For a typical monitoring system of a wind turbine vibrations are recorded from 6 channels, each approximately every 10 min. Therefore, the number of all raw signals acquired in 1 day is: 6 channels * (60 min/10 min) * 24 h = 864 raw signals/day. Since storage of a single 10-s raw vibration signal uses more disk space than a trend of a signal feature over a year, typically only 1 signal per day is stored permanently (or even less). Consequently, as shown in the given example, the probability of virtually blind selection of raw signals corresponding to the smallest fluctuation of operational parameters via fixed data storage configuration is 6/864 $\approx$ 0.7%, which might extremely deteriorate the entire diagnostic process. Therefore, the proper selection of signals for calculating signal features and occasionally for permanent storage of raw vibration data is so crucial. Only "the best" signals must be chosen.

4.2.2 Selection Based on Operational States

As described in the Chap. 2, the data acquisition process in condition monitoring systems is based on operational states. The operational state of a machine is defined as a range of operational points in a machine enabling referential analysis with stored baseline data. In practice, an operational point is defined by measurement values of process channels such as speed, load, pressure, temperature, etc., usually called "machine process parameters". The described data selection procedure for wind turbines serves as an auxiliary tool to a well-known concept of definition of machine states. Typically, rotational speed of a generator and output power are used for the definition of states. In industrial applications machine operational states are based either on nominal data, standards or on reliable historical data.

Nevertheless, in many commercial monitoring systems, definition of operational states is based solely on a single operational parameter. Typically, it is either generator power or rotational speed of a generator shaft. Apart from (often decisive) economic reason, a single-parameter-based definition is most likely chosen due to undesirable

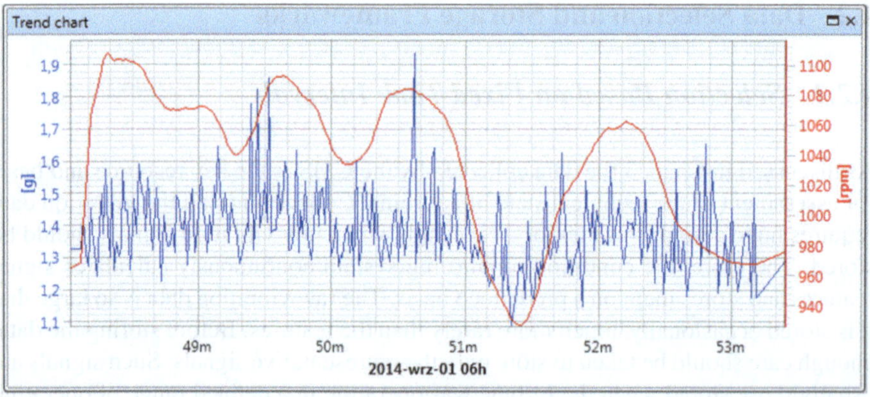

Fig. 4.2 Variability of RMS of bearing housing vibration signals as a function of varying speed

complexity of multi-channel operational states. Moreover, implementation of large distributed CMS systems is still a relatively new field of engineering. Therefore, some unsuitable well-established solutions dedicated to relatively simple machinery have been inevitably transferred to modern systems, including fixed data storage conditions mentioned above.

Due to possible significant fluctuation of wind speed a wind turbine works under extremely variable operational conditions. They influence rotational speed of blades which—in turn—directly influences produced power. As a result, a wind turbine is characterized by highly variable dynamic characteristics that could easily change values of amplitudes of vibration-based health indicators. Figure 4.2 illustrates an exemplary RMS estimator whose variability due to variation of operational conditions (here: the rotational speed of the generator) reaches 90%. It is worth mentioning that during the presented 5 min period the speed changed only by 150 RPMs.

In order to limit such large variabilities of health indicators, it is recommended to introduce operational states (apart from averaging described i.e. in [6]). The definition of operational states enables to deduct whether a change in magnitude of health indicator is generated simply by varying operational conditions or by deteriorating technical condition of a machine. In commercial condition monitoring systems the definition of operational states is done on the basis of one or more operational parameters. Typically, these may include:

- rotational speed of a generator,
- generator active power,
- rotation of a tower (i.e. azimuth drive activation),
- wind speed.

While in case of older systems the definition of operational states was done on the basis of a single snapshot-value of just one operational parameter modern CMS systems are able to track many process parameters even during signal acquisition.

A failure to meet the operating state condition might result in signal registrations during the tower rotation and it introduces a significant amount of noise leading to possible false alarms. On the other hand, a failure to track the operational parameter during the data acquisition may result in a partial registration of vibration signals when the machine is no longer in the required state, since during long measurements a wind turbine often changes values of operating parameters. If the required data sampling should be 60 s (e.g. for a slow rolling main bearing or planetary gearbox), it is not uncommon for the power of the turbine to be changed so much that the operational state is not the same as it was a moment ago.

Due to the fact that warning and alarm thresholds are determined independently for each operating state, it is a good practice to define as few operational states as possible (typically up to three). In most cases related to monitoring of a wind turbine status two operational states are implemented.

4.3 Process Parameters Validation

Validation of process parameters means continuous reading of available machine process parameters. The data is nonstop analyzed to prove its validity and coherence. This task can be accomplished in two steps. First, each variable is analyzed independently from the others. Then, the multi-dimensional analysis is performed (when more than a single variable is considered).

4.3.1 Single Process Parameters Validation

The first step—the single variable validation—is a method which analyses each process variable independently. Predefined maximum and minimum thresholds are set on the current values of process parameters. For example, given a wind turbine operating with speed up to 1500 rpm, the absolute limits might be set at zero and 2000 rpm as lower and upper boundary, respectively. Reading outside the range is impossible from the physical point of view and—if experienced—will signal an error in the data. Such limits should be set for all the process parameters.

Apart from a direct parameter value, certain process parameters are never stable, but vary even at a stable operational point. Parameters such as load in the case of underground conveyors [7] or wind speed [8], might be required to change its value within certain time, since it might be assumed that theses physical quantities must express certain variability.

A third test which is linked to the second one is the implementation of constrains for the maximum allowable instantaneous change of process values, i.e. to prohibit sudden changes. Just as, in reality, within a fraction of a second, the wind will not suddenly die in the next sample if it was 25 m/s in the previous one, or the generator rotational speed will not change by 1000 rpm from one reading to another. One powerful technique to achieve such constrains is to set a threshold on a maximum value of derivative of time series. One should remember that calculating derivatives is vulnerable to noise and a robust method should be applied. The discussed test should have limits much larger than typical noise values experienced during normal operation.

4.3.2 Multi-dimensional Process Parameters Validation

The multi-dimensional validation exploits the physical relationships between the process parameters. If a turbine is in a certain state (for example the nominal power is 1.5 MW), then the other process parameters must stay within certain limits (even very wide), for instance the speed must be in a range from 1000 to 1500 rpm. The development of such multi-dimensional validation rules can be done manually using the a priori knowledge about the turbine or the historical data. Often, the rules are not reflexive. For example, in the case of an examined stall-controlled wind turbine at instantaneous 1500 rpm speed the power might experience large fluctuations because it is a function of wind as well, as marked in Fig. 4.3. In this case, the output power decreased from 680 to 114 kW in less than 4 min. In the same time the wind speed was varying from 14 to 7 m/s, but the turbine had constant rotational speed equal to 1500 rpm as it is typical of stall controlled machines.

Therefore, it seems most reasonable to deduce required dependencies taking into account all available process parameters which influence one another. Such validation requires sufficiently long historical data and a well-organized definition of machine operational states (including run-ups and run-downs detection). This type of validation using rather historical data than physical relationships is a very interesting field for machine learning algorithms. With large amounts of available data, the goal of the analysis is reduced from fault detection to much simpler anomaly detection.

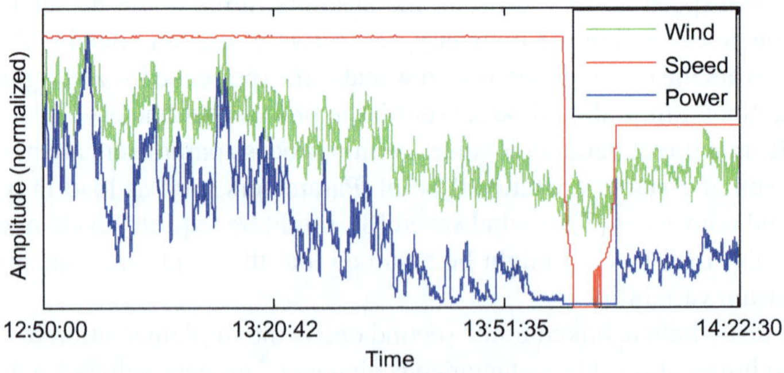

Fig. 4.3 Example of power and wind fluctuation at nominal speed (reprinted from [4] with permission from Elsevier)

Up till now, all the validation tests have used instantaneous values of the parameters only. They have not taken into account the dynamics of the process. Another technique which might be applied at this stage of validation is a time delay parameter. Only if selected rules are violated for at least a minimum time the data is marked invalid. It can also be a field of development and application for more complex dynamic models.

4.4 Vibration Signals Validation

4.4.1 Signal Selection Criteria

Let us first consider common errors encountered during the acquisition of vibration signals. Some examples will be used to build the intuition about signal validity. At first, the problem will be described from the "common sense" point of view. Next, several examples of real vibration signals are presented.

A correct vibration signal is the one which truly represents the machine behavior. Generally, dynamic behavior of a machine should follow expectations regarding the frequency content present in the signal, with some level of accuracy. To a large extent, a field engineer is able to verify the correctness of a vibration signal at a glance. However, a verbal description of the mental process while performing such an immediate validation is far more difficult than a simple, though very imprecise: "it just looks fine". In order to create an algorithm being able to distinguish valid signals, a set of more precise rules should be proposed. The following list attempts to describe verbally some characteristics of a valid vibration signal (from a continuously running machine):

- mandatory visual continuity,
- required certain complexity (as opposed to a computer-generated sine wave),
- rational amplitude levels,
- imperceptible quantization (for sufficiently long time period),
- sufficient sharpness of the time waveform shape (due to expectations of high frequency components),
- sudden signal changes present only to a degree allowable by the machine real behavior,
- expected mean value accuracy (zero in case of acceleration).

Likewise, an incorrect signal is defined as one which does not represent the machine behavior as expected. The reason for a signal being incorrect is not always known. There can be a number of potential reasons, e.g. a hardware error, a signal processing software error, sudden random impacts or a combination of those may occur. However, the exact source of an error is not of a major concern, especially if the event was of a single occurrence. The most important task is to identify and

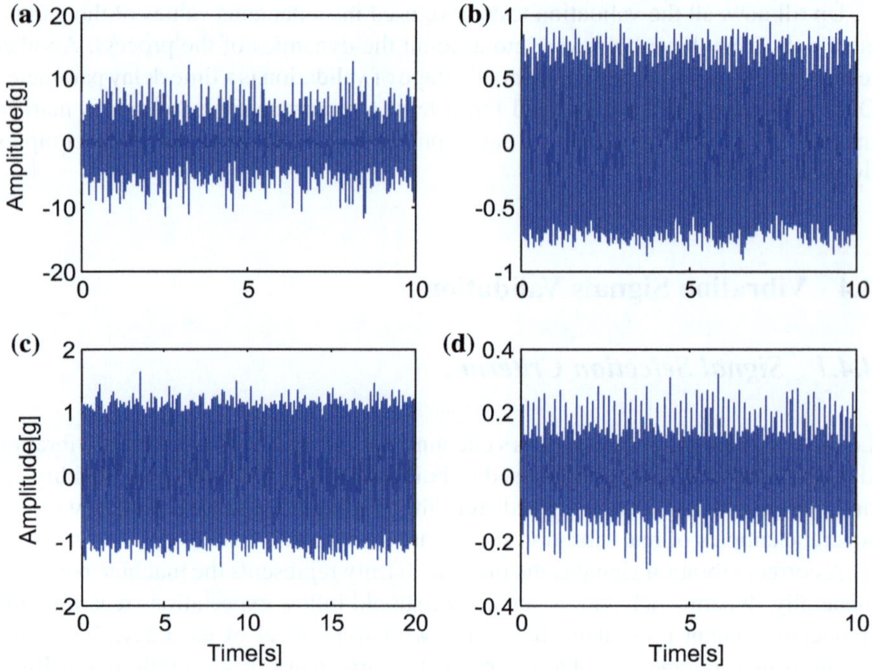

Fig. 4.4 Examples of correctly recorded vibration signals (reprinted from [5] with permission from Elsevier)

reject invalid data before it is used for feature calculation and the data-based analysis is stored permanently in the system while the original raw signal is deleted.

A number of real valid vibration signals from wind turbines is presented in Fig. 4.4. Then Fig. 4.5 illustrates a group of real (i.e. recorded on a real machine) signals which are obviously invalid. Signals (a) through (j) are used in the further text as the examples. All of these signals are real signals sampled at 25 kHz. It is worth mentioning that raw vibration signals may include the tacho signal information, i.e. samples acquired at a certain phase angle were marked. The tacho signal marks are not displayed in Figs. 4.4 and 4.5 since here the main focus is on the vibration signal only. The discussion of speed validation is not included here as it is considered the process parameter and was discussed in the Chap. 4.3.

After a quick visual examination one can state that signals (a–d) from Fig. 4.4 fulfill the requirements for a correctly recorded signal listed above. Each and every data does not show any sudden changes in the signal level. All data exhibits a relatively complex content and does not show a visible quantization. In contrast, the data displayed in Fig. 4.5 suffers from sudden changes of amplitude. It exhibits discontinuity, asymmetry, apparent quantization and relatively rapid and random signal pattern changes. At this point, it is important to state that correctness assessment is sometimes not at-a-glance operation since some signals may have real causes. For

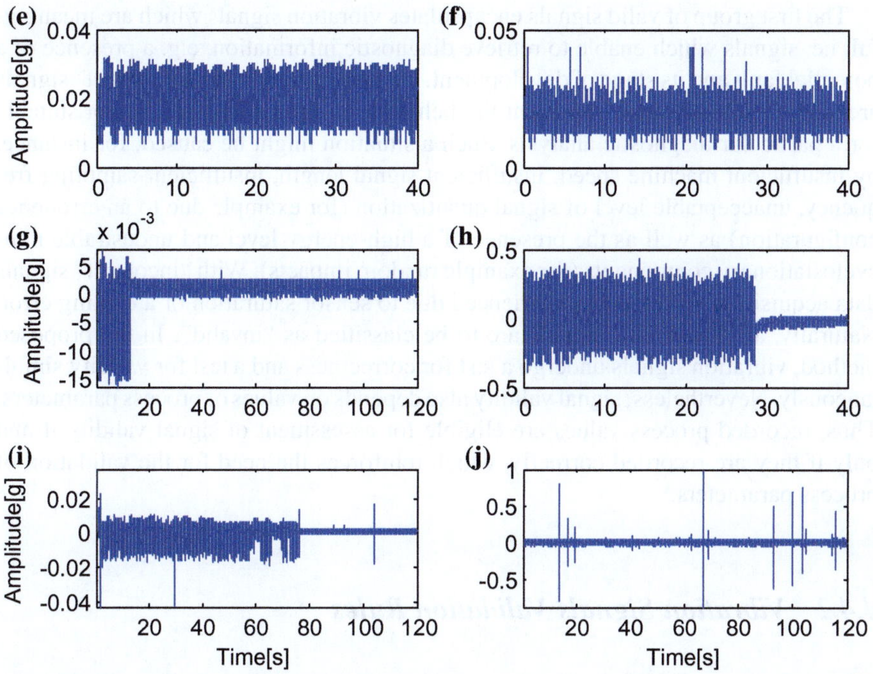

Fig. 4.5 Examples of incorrectly recorded vibration signals (reprinted from [5] with permission from Elsevier)

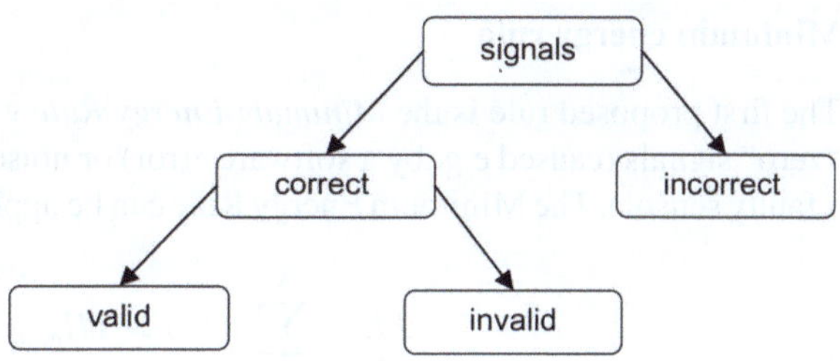

Fig. 4.6 Classification of vibration signals (reprinted from [4] with permission from Elsevier)

example, such impulses as illustrated in Fig. 4.5j may either represent real impulse responses with a relatively high damping or stand for single large positive and negative values generated arbitrary by hardware or software errors. Nevertheless, such signals must not be discarded and they require additional rules in order to verify their correctness.

Having examined the examples of the signals in Figs. 4.4 and 4.5, it is necessary to distinguish more precisely between different classes of signals. Jablonski and Barszcz [4] proposed a distinction between correct, incorrect, valid and invalid signals. Correct vibration signals might be classified as valid or invalid, as illustrated in Fig. 4.6.

The first group of valid signals encapsulates vibration signals which are meaningful, i.e. signals which enable to retrieve diagnostic information, e.g. a presence of a potential fault and its stage of development. On the contrary, “invalid correct” signals are such which still truly represent the behavior of a machine, but are not suitable for a particular diagnostic analysis. Such a situation might be caused, for instance, by insufficient machine speed, insufficient signal length, insufficient sampling frequency, unacceptable level of signal quantization (for example due to an erroneous configuration) as well as the presence of a high-energy level and undesirable non-cyclostationary components (for example random impacts). With “incorrect” signals data acquisition faults were experienced due to sensor saturation or a cabling error. Naturally, all “incorrect” signals are to be classified as “invalid”. In the proposed method, vibration signals undergo a test for correctness and a test for validity simultaneously. Nevertheless, signal validity also depends on values of process parameters. Thus, recorded process values are eligible for assessment of signal validity if and only if they are recorded correctly, which reinforces the need for the validation of process parameters.

4.4.2 *Vibration Signals Validation Rules*

This chapter presents in detail proposals for validation rules. Each method is presented with mathematical formulas. The chapter also includes the case study presenting results of applying these methods for a set of real signals.

Minimum energy rule

The first proposed rule is the *Minimum Energy Rule* which is intended to recognize “zero” signals (caused e.g. by a software error) or noise-level signals (caused e.g. by a faulty sensor). The Minimum Energy Rule can be applied either from the definition:

$$E = \sum_{1}^{N} x^2(n) > TH_{Min_E} \tag{4.1}$$

where:

$x(n)$ time signal amplitudes
N number of signal samples
E signal energy
TH_{Min_E} minimum energy threshold

or via a root-mean-square (quadratic mean) estimator:

$$RMS = \sqrt{\frac{1}{N} \sum_{1}^{N} x^2(n)} > TH_{MinRMS} \tag{4.2}$$

The rule is simple, but powerful. It states that any vibration signal which was recorded from a machine with the rotational speed above a certain limit should carry at least minimum predefined energy. The practical reason is also straightforward. The rule will be violated e.g. when a vibration sensor is detached from the machine.

The rule performance as well as threshold settings are straightforward, since the difference between noise-level energy and real signal energy is of at least a few orders of magnitude for industrial machines. The fulfillment of this requirement may be treated as a *sine qua non* condition for further validation.

Amplitude range dynamics rule

In distributed systems of monitoring and diagnosis where hundreds of vibration channels are installed with default channel range settings, it is of utmost importance to adjust each vibration channel range properly. Otherwise, sensor saturation or unacceptable level of signal quantization might occur. For instance, for common CMS setup with 6 sensors, the average level of vibrations at the main bearing (lowest speed) is more than order of magnitude lower than average level of vibrations at generator bearings (highest speed). Typically, there should be a guideline for a given type of a turbine and a given CMS how the proper channel ranges should be configured.

Having configured the system properly, nominal data should cover 15–20% of a total channel range. Then, a relatively small coverage of range might inform about a sensor failure [9]. With this level of dynamics readouts have sufficient dynamics, even with ADC of lower number of bits (14 or 16). Yet, in case of the increase in the signal value the converter will have sufficient room before saturation occurs. This is a very unfortunate situation, as a signal cannot be tracked anymore.

N-point rule

At the beginning of this chapter it has been stated that a human eye may be a significant tool for signal validation and that the need for automatic validation refers only to large databases in which majority of signals is stored temporarily, hardly ever being displayed on a screen. However, there is a group of signals which are insensitive to visual assessment. Such signals, for instance, contain a number of consecutive samples with exactly equal amplitudes which is the basis for the proposed "N-point" rule. In the case of hardware with typical dynamics (e.g. 16 bit AD converters which are still common in industry), such a situation may be controlled. The method can be presented as the answer to the following question: how many consecutive signal values shall be allowed to have exactly the same values for a correctly recorded vibration signal (for instance 10 s)? This seemingly easy problem is by no means trivial, since apart from the signal itself it depends on both the sampling frequency and the AD characteristics.

According to the signal theory, a single sinusoidal signal after digitization with a particular sampling frequency fs may contain only two consecutive samples with exactly the same amplitudes as shown in Fig. 4.7.

On the other hand, for a real vibration signal (which should be rather modelled as the stochastic process than a sine function), the possibility of such occurrence

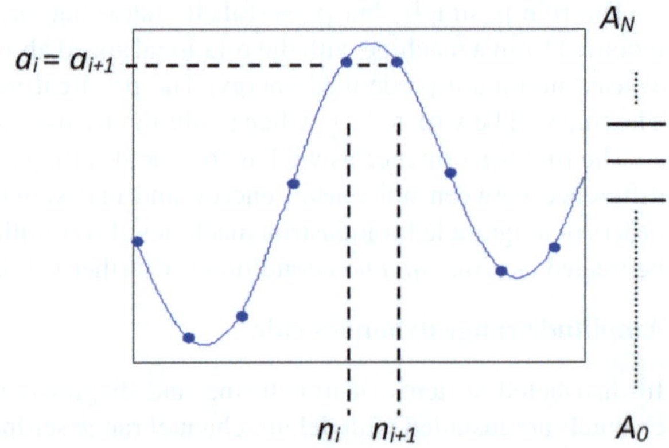

Fig. 4.7 Illustration of two consecutive samples with exactly the same amplitudes for quantization levels from A_0 to A_N (reprinted from [5] with permission from Elsevier)

Table 4.2 The number of n consecutive samples with exactly the same amplitudes for a population of one thousand correctly recorded 10 s vibration signals (reprinted from [5] with permission from Elsevier)

N	No. of instances
2	715,473
3	22,092
4	959
5	46
6	0
7	0

is a random function with the probability close to the reciprocal of the number of possible amplitude values. It is apparent that the probability depends on the number of bits of the AD converter. Note that for a larger number of consecutive samples with exactly the same amplitudes the probability is multiplied by the same factor. Therefore, it decreases significantly.

The empirical research has shown that for correctly recorded signals of vibration acceleration (with 16 bit ADC), even four consecutive equal amplitudes can be recorded. The main reason for this fact is that real signals are not sine functions, but should be rather analyzed as random series. In practice, the increased number of samples having the same value may be a result of general inevitable imperfection of the electronic hardware or the saturation of electronic elements. There was a test conducted on a population of one thousand 10 s vibration signals, and the results are presented in Table 4.2. The sampling frequency was 24 kHz and 16 bit AD converter was used.

From Table 4.2 one may conclude that a reasonable threshold may be selected. Conducting analogous investigation for the data from Figs. 4.4 and 4.5, Table 4.3 presents the results of the *N*-point rule for the exemplary data. Clearly, the correct data differs significantly from the incorrect data according to the *N*-point rule.

For the valid data from Table 4.3, no instance has experienced more than three consecutive points with the same amplitudes, whereas for the invalid data, all the signals contained instances of at least ten consecutive points with exactly the same amplitudes.

Table 4.3 The number of *n* consecutive samples with exactly the same amplitudes for signals from Figs. 4.4 and 4.5 (reprinted from [5] with permission from Elsevier)

	N	2	3	4	5	6	7	8
Valid	a	385	3	0	0	0	0	0
	b	535	3	0	0	0	0	0
	c	950	3	0	0	0	0	0
	d	2727	25	0	0	0	0	0
Invalid	e	230,394	93,109	44,950	30,172	21,019	15,660	12,101
	f	482,553	186,876	80,681	46,839	28,183	18,676	1331
	g	1,389,512	572,896	268,046	161,985	99,246	64,086	28,590
	h	1,441,196	582,105	270,334	162,619	99,425	64,132	10
	I	2,398,516	1,018,785	492,587	286,862	171,997	108,560	27,692
	j	3,700,854	1,761,656	976,451	616,721	400,965	273,666	120,695

Formally, the N-point rule may be written as follows:

$$(\forall x{\in}X)(\neg\exists\, x \in X)\big(\forall_{i=i,i+1,\ldots,i+N-1}(x_i = x_{i+1})\big) \tag{4.3}$$

where:

X set of signal values
x single signal value
N N-point rule coefficient

and can be interpreted as: "For all the samples in the signal x, there is no sample for which N consecutive samples have the same value."

A final remark concerning the N-point rule is that the validation should allow consecutive equal amplitudes in case when the signal reaches the measurement range boundaries. Such a condition is necessary because a machine which experiences a large fault may generate relatively large vibrations which ultimately saturate the sensor. In such cases the extreme amplitude may be captured by the hardware for a fraction of a second which will result in a high number of consecutive samples of the same, maximum (or minimum) value.

U-point rule

One of the initial requirements towards vibration signals was that they should exhibit "required certain complexity". This condition can be expressed as a requirement for a signal to use a number of unique amplitude values in a signal for its validation, and it is called by the author the "U-point" rule (from "unique"). After analog—digital conversion real values are changed into binary coded numbers. Putting aside repetitions in a single time series the number of unique values should be above a threshold. The number of unique points can be used as a complementary tool to a signal amplitude range dynamics rule, as illustrated in Fig. 4.8.

As illustrated in Fig. 4.8 a signal might cover sufficient sensor dynamics, i.e. range at expected level. However, it might be still incorrect, like "sig2" which is a simple square wave and is not a correct vibration signal. Please note that the "sig2" is characterized by the very same range dynamics as the "sig1". On the other hand, these

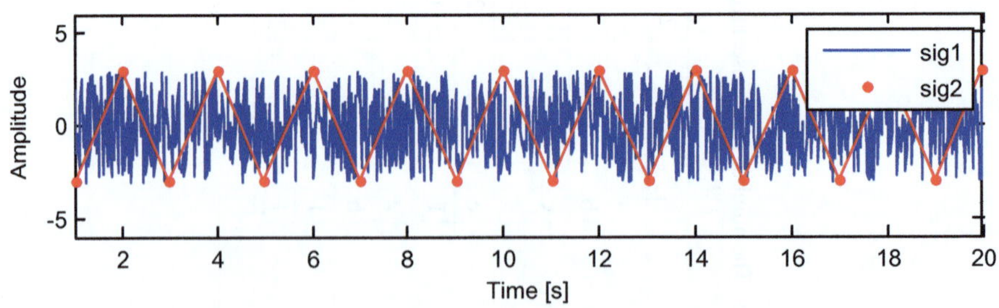

Fig. 4.8 Comparison between range dynamics and number of unique points of a signal (reprinted from [4] with permission from Elsevier)

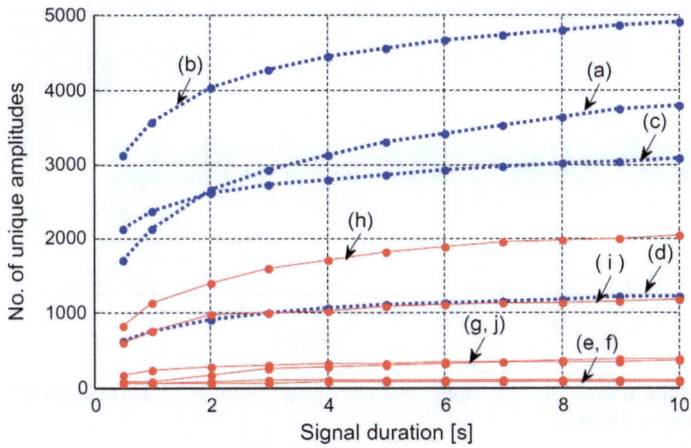

Fig. 4.9 Number of unique samples as a function of number of total samples (signal duration) for signals (**a**–**j**) (reprinted from [5] with permission from Elsevier)

signals are very different in terms of the number of unique value points, e.g. "sig1" has 1000 unique points whereas "sig2" only 2 unique points. In order to compare all the presented signals Fig. 4.9 illustrates calculated numbers of unique points from signals a-j as a function of the signal duration.

As illustrated in Fig. 4.9 for standard measurements with sampling frequency 25 kHz and the 16 bit ADC converter, a threshold of 2500–3000 unique points could be established with the capability of detection of more than half of invalid signals.

Zero-crossing rule

Acceleration signals from machinery oscillate around the zero value and the averaged vibration value is also zero (as long as the machine isn't moving which is not the case for a wind turbine). Therefore, another rule can be proposed which will detect abnormal behavior as can be seen in Fig. 4.10. The proposed "zero-point rule" (or "Z-point") requires that there is a maximum number of consecutive samples without crossing the zero value. Figure 4.10 presents a real vibration signal violating the Z-point rule.

The situation as presented in Fig. 4.10 may be caused by a phenomenon caused by saturation of an accelerometer. Such a case may be a result of an external impact. The "Z-point" algorithm has two steps:

(a) calculation of a maximum number of consecutive samples larger than or lower than zero,
(b) comparison of the calculated number with the maximum allowable threshold TH.

The algorithm can be written down in Matlab as follows in a single line:

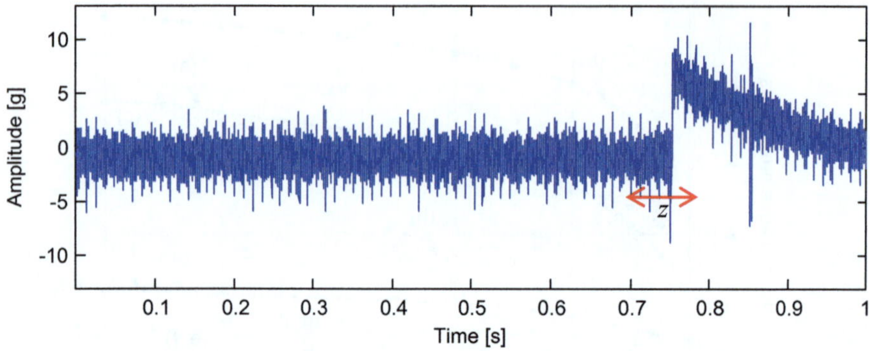

Fig. 4.10 Typical example of a faulty vibration signal violating *Z*-rule (reprinted from [4] with permission from Elsevier)

$$\begin{aligned}&\texttt{max}(\texttt{max}(\texttt{diff}([0,\ \texttt{find}(\sim [0, \texttt{x} < 0,\ 0])]) - 1),\\&\qquad \texttt{max}(\texttt{diff}([0, \texttt{x} > 0,\ 0])]) - 1)) < \texttt{TH}\end{aligned} \tag{4.4}$$

where:

x time series
diff a derivative (difference between consecutive samples)
find standard indexing function
TH empirical scalar threshold value.

The notion [0, x < 0, 0] means adding zeros at the beginning and at the end of a time vector in order to handle the time series boundaries. Violation of the Z-point rule classifies a signal as invalid.

The exact value of the *TH* level depends very much on the parameters of a measurement system. The most important ones are: sampling frequency, sensor range and sensor sensitivity. In case of the investigated acceleration signals, it was observed that correct signals had a *TH* no larger than 200, while for the incorrect signals there were even ten times higher values.

4.4.3 Signal Stationarity Validation

As it was proposed in the Sect. 4.4.1, signals can be categorized as correct or incorrect due to the correctness of sensor operation and the data acquisition path. Moreover, correct signals may still be valid or invalid. The latter happens when a signal represents the physical behavior of a machine correctly, yet it is unsuitable for data analysis. This may happen due to the unacceptable level of change in a signal content. Another case is when the turbine is stopped. A sudden change in a wind turbine load will result in a large change in vibration level. Such transient states are sometimes a valuable source of data (e.g. large turbomachinery), but are not used in wind turbines. Therefore, signals suitable for analysis should be as stationary as possible. The signal stationarity can be checked in a number of ways.

Signal fragmentation techniques

Taking into account signal stationarity requirement, the simplest method could be a comparison of consecutive fragments of a signal. The method will detect when the level of change is larger than the assumed level, as proposed by Jablonski and Barszcz [10]. In machinery vibration analysis one assumes that a machine produces vibrations characterized by existence of at least one full cycle with respect to which the machinery operation is periodic, as it is the fundamental assumption of cyclostationarity [11]. Figure 4.11 presents the examples of non-stationary vibration signals which violate this assumption.

The pulse response in Fig. 4.11 (left) does not represent any finite cycle so the signal is not suitable for frequency analysis, nor is it useful for general condition monitoring and it can be considered as invalid. Although it may happen that some fragmented (or partial) cycles will be included within the acquired vibration signal, such a signal should be rejected as it affects trend analyses, like for instance peak-to-peak value. Such a fragmentation may take place if the signal sampling time is several seconds and the longest cycle (which is the planetary gearbox) can be even several minutes. The issue of detection of such signals is different from detecting a pulse response, as described in the autocorrelation-based works [12].

The status of an invalid signal may also be caused by a rapid change in the operational parameter, as depicted in Fig. 4.11 (right). As proposed at the beginning of this chapter, such invalid signals can be detected by means of a simple comparison of signal fragments, see Fig. 4.12. The intervals of significantly different characteristics are marked with letters a, b, c, and d.

The proposed method requires two parameters, namely the selection of length of a signal fragment and the selection of a signal feature to be compared. It seems that the simplest feature which is the peak-to-peak (PP) amplitude is sufficient and at the same time very simple to calculate. The selection of length can be chosen in several ways. Three methods will be presented here: a fixed fragment duration, a fixed number of

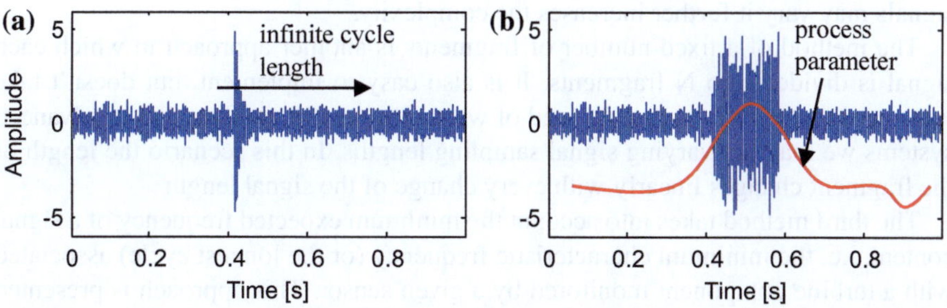

Fig. 4.11 Examples of non-stationary vibration signals violating fundamental assumptions of cyclostationarity (reprinted from [4] with permission from Elsevier)

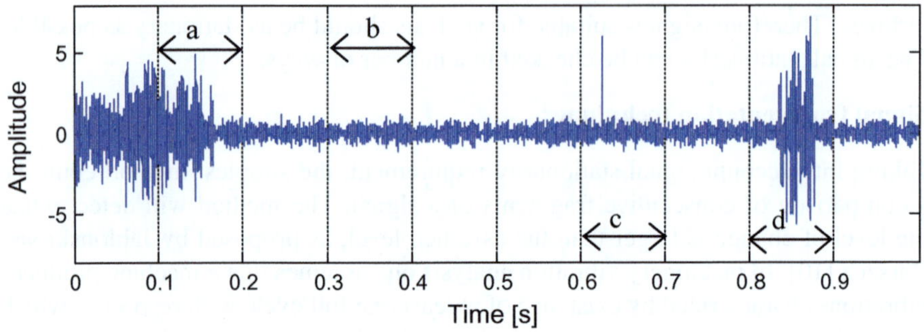

Fig. 4.12 Illustration of a comparison of signal fragments (reprinted from [4] with permission from Elsevier)

Table 4.4 Methods of signal fragmentation

Method	Advantages	Disadvantages
Fixed time	Simplicity	Will not give good results for variable rotational speed machines
Fixed number of fragments	Simplicity	Will not give good results for varying length signals
Lowest expected cyclic frequency	Proper representation of machine behavior	More complexity, will require additional parameters

fragments and a fragment containing the lowest expected characteristic frequency. Table 4.4 presents these methods and lists their advantages and disadvantages.

The method of a fixed time fragment seems at the first glance to be the obvious choice. It is effortless and simple to implement. A vibration signal is divided into fixed time intervals, e.g. 500 ms. Next, for each fragment the feature (PP as proposed above) is calculated. Finally, variability of PPs for all the fragments is checked and compared with a threshold. The main drawback of the method is that wind turbines have varying rotational speed. Consequently, all the characteristic frequencies are speed dependent. Since in an advanced system data acquisition times of vibration signals may vary it further increases the complexity.

The method of a fixed number of fragments is another approach in which each signal is divided into N fragments. It is also easy to implement, but doesn't take into account varying rotational speed of wind turbines and the fact that in advanced systems we may get varying signal sampling lengths. In this scenario the length of the fragment changes linearly with every change of the signal length

The third method takes into account the minimum expected frequency of a signal content, i.e. the minimum characteristic frequency (or the longest cycle) associated with a turbine component monitored by a given sensor. This approach is presented in Fig. 4.13. This method is more complex to implement, as it should be adapted to

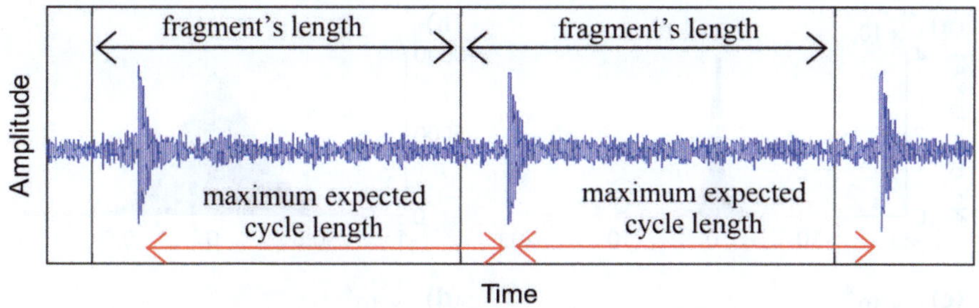

Fig. 4.13 Lowest expected cyclic frequency signal fragmentation method (reprinted from [4] with permission from Elsevier)

a machine drivetrain configuration. It can be adapted to varying rotational speed and changes in signal length, though.

Since the maximum cycle length is a function of machine speed, this method requires recalculation of the fragment lengths for every signal. The implementation of this method is simpler if signals are first resampled and thus contain the equal number of samples per revolution. Then the fragment length is to be calculated using a known ratio of the lowest characteristic frequency to the reference signal. When a raw signal is to be analyzed, the fragment length needs to be calculated with respect to the slowest machine cycle. This will need recalculation of fragment lengths for each sensor separately because generally the vibration contents strongly depend on the localization of a sensor and different components are monitored by different sensors.

4.4.4 Probability Distribution Function Fit

Previous chapters presented the methods of vibration signal validation using the amplitude values and parameters derived from raw time signals. Another possible approach is to investigate the statistical parameters of raw signals.

The first step in this approach is to test the vibration signal for how close it is to the Gaussian distribution. During the research, the author has investigated distributions of a large number of raw vibration signals in terms of:

- location parameter μ
- scale parameter σ.

The results for the signals presented in Fig. 4.4 (correct ones) and Fig. 4.5 (incorrect ones) will be discussed. The first step of the analysis was the calculation of the best fit of the probability density function (PDF) given by the Eq. (4.5) of Gaussian signal amplitudes:

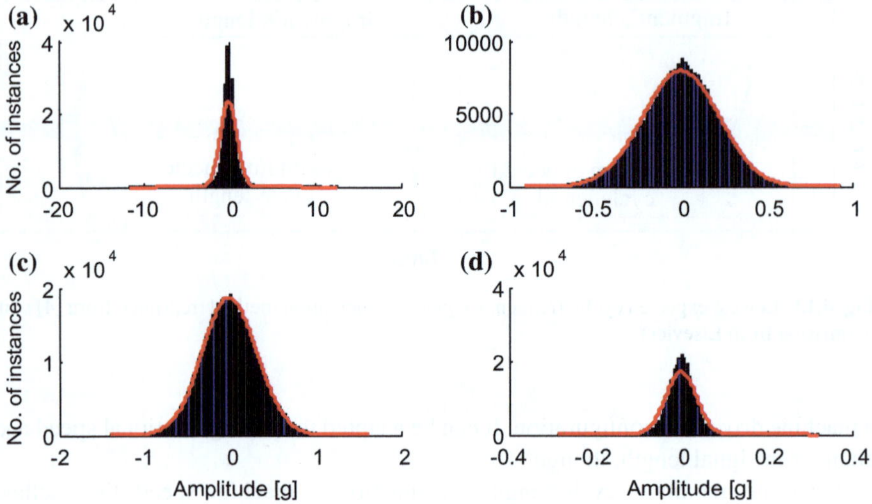

Fig. 4.14 Histograms and PDF Gaussian fits corresponding to data from Fig. 4.4 (valid signals) (reprinted from [5] with permission from Elsevier)

$$f(x|\mu, \sigma) = \frac{1}{\sigma\sqrt{2\pi}} e^{\frac{-(x-\mu)^2}{2\sigma^2}} \tag{4.5}$$

Firstly, the location parameter μ must be calculated. For acceleration signals, this parameter is the offset value, and it should be close to zero. As shown in the examples in Fig. 4.5e, f, this parameter can detect signals for which the zero offset value is not satisfied. In general, for the Gaussian distribution the parameter μ is the mean value. As such it is easy to calculate and does not require PDF calculation.

Secondly, the scale parameter σ, depicts the flatness of the distribution. The study has shown that σ is amplitude dependent and does not seem to carry valuable information about the signal validity.

The next step of probability distribution fit can be started with visual investigation of the signal histograms and its comparison with the PDFs of the Gaussian distribution fitted to the signal data. Figures 4.14 and 4.15 show histograms with corresponding PDFs for signals introduced in Figs. 4.4 and 4.5. The visual inspection immediately presents very different characteristics of these two groups of signals. The valid signals appear to be much closer to the Gaussian distribution while the invalid ones have much different distributions. Clearly, the assumption about Gaussianity of invalid signals can not be held true.

Please note, that data histogram strongly relies on the predefined number of bins. The number of 100 bins was assumed for this research, as it can be well presented for visual inspection and it was also applied for numerical analyses.

The visual inspection can only be the first step of the analysis. For an automated method a numerical test should be proposed. There are several "goodness-of-the-

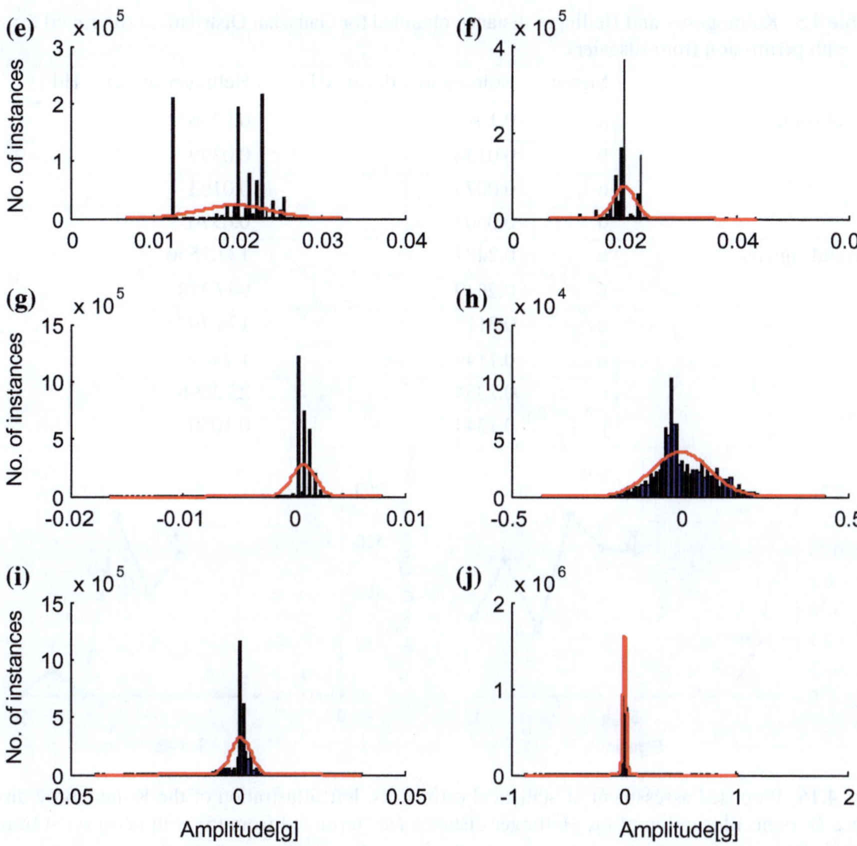

Fig. 4.15 Histograms and PDF Gaussian fits corresponding to Fig. 4.5 (invalid signals) (reprinted from [5] with permission from Elsevier)

fit" tests available [13]. The two tests, namely the Hellinger distance [14] and the Kolmogorov test were applied.

The Hellinger distance is defined as a measure of divergence between the empirical distribution P and a model distribution Q and is calculated as given in (4.6):

$$Hd = \sqrt{\frac{\sum_{t=1}^{t_{max}} (P(x_t) - Q(x_t))^2}{n}} \tag{4.6}$$

where:

$P(x_t)$ is empirical distribution,
$Q(x_t)$ is fitted distribution.

The smaller the value of Hd, the bigger the similarity between $Q(x_t)$ and $P(x_t)$ is.

The Kolmogorov distance [15] is given by Eq. (4.7):

Table 4.5 Kolmogorov and Hellinger distance obtained for Gaussian Distribution (reprinted from [5] with permission from Elsevier)

	Signal	Kolmogorov distance D	Hellinger distance Hd
Valid signals	a	0.1006	0.0476
	b	0.0130	0.0399
	c	0.0073	0.0163
	d	0.0507	0.6970
Invalid signals	e	0.2487	137.2556
	f	0.2549	99.6378
	g	0.3112	176.7071
	h	0.1149	1.1407
	i	0.2354	25.2696
	j	0.1344	0.1050

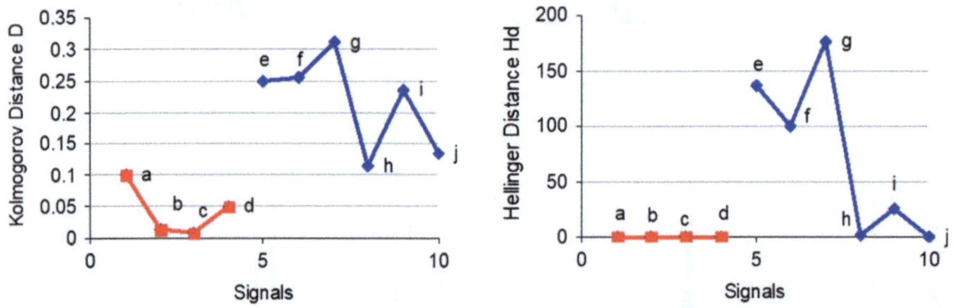

Fig. 4.16 Proposed assessment of statistical estimators, left: illustration of the Kolmogorov distance *D*, right: illustration of the Hellinger distance *Hd* (reprinted from [5] with permission from Elsevier)

$$D = sup_{t \in R}|P(x_t) - Q(x_t)| \tag{4.7}$$

and is a distance between the empirical distribution $P(x_t)$ and a fitted distribution $Q(x_t)$. As in the previous case, the smaller the value *D*, the bigger the similarity is. The calculated values of Kolmogorov distance *D* and Hellinger distance *Hd* for all the signals from Figs. 4.4 and 4.5 are shown in Table 4.5 and in Fig. 4.16. The same letters for signal descriptions were used.

From Table 4.5 and Fig. 4.16, it can be observed that there are large differences between values of statistical estimators for the valid signals and most of the invalid ones. However, there are invalid signals (h, j) which have values of *Hd* and *Kd* similar to the valid ones. Thus, the proposed statistical data validation may be used as an auxiliary tool only.

It can also be stated that for the Kolmogorov distance valid signals and invalid signals are separated, yet the distance between them is relatively small (it is less than the *Kd* variation within each group). For the Hellinger distance the data sets are not fully separated, *Hd* values for 4 signals (signals e, f, g, i) are over ten times greater than the largest estimator value from the valid data signals (d).

4.5 Complete Data Validation Algorithm

After presenting all the validation methods, it is now possible to propose the complete data validation algorithm. Figure 4.17 presents the flowchart of the proposed data acquisition and the validation algorithm which includes the steps proposed throughout the Chap. 4. It is assumed that the flowchart describes a single recording period which can be one day. During this period the operation of the system is continuous. The flowchart is shown for a single operational state.

The flowcharts symbols are:

- N_{i} required number of vibration signals to be permanently stored corresponding to operational state i,
- n_{i} number of stored vibration signals corresponding to operational state i,
- S_i set of all stored vibration signals corresponding to operational state i characterized by the value of the stationarity criterion s_i,
- s_i any vibration signal within S_i characterized by the value of the stationarity criterion,
- s current vibration signal characterized by the value of the stationarity criterion.

As presented in Fig. 4.17, first the process parameters are validated. For this step defined rules for a single channel and a multi-channel should be used. Next, the system checks all the PVs until the turbine enters a defined operational state and starts recording a raw vibration sample (flagged as belonging to this operational state). During the acquisition process 4 actions should be continuously performed, namely: validation of process parameters, check whether the turbine stays within the current operational state, validation of permissible fluctuations of process parameters and validation of vibration signal (several tests). If any of the abovementioned conditions fails the data is deleted and no features are calculated. After the successful acquisition of a vibration signal and corresponding process parameters, a defined stationarity coefficient is calculated from the process parameters. All defined signal features are calculated and written into a trend database as well. Storage of raw vibration signals is repeated for each acquisition period until a required number of raw vibration signals in a given operational state is met. With all the incoming data, all the steps are repeated. If the stationarity coefficient of the current time signal is smaller than the stationarity coefficient of any stored samples the least stationary raw time signal (i.e. with the largest value of a stationarity criterion) is replaced by the current sample. As a result at the end of each acquisition period (e.g. day), the "most stationary" raw vibration signals are stored in the database and the signal features for all the valid samples are written as trends. The flowchart from Fig. 4.17 is shown for a single operational state only and it has to be implemented separately for each defined operational state. Therefore, the best data is stored for every operational state.

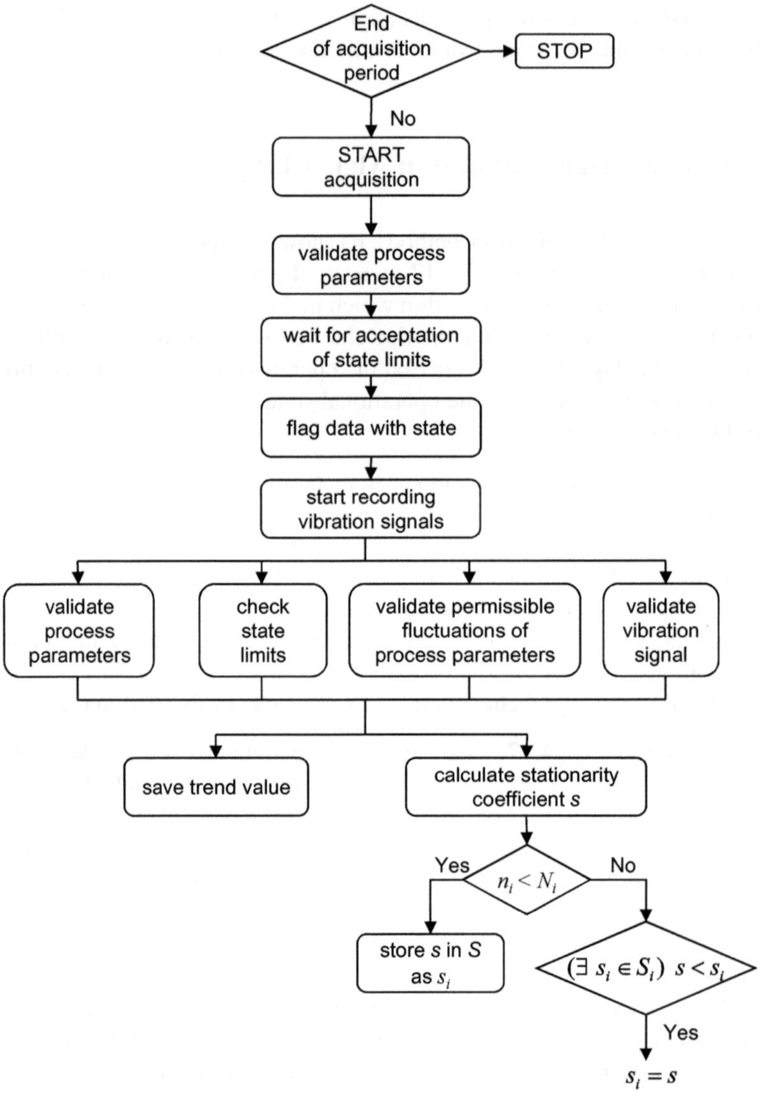

Fig. 4.17 A proposed data selection flowchart (reprinted from [4] with permission from Elsevier)

The implementation of the proposed algorithm will result in very high quality data, and the calculated features can be compared with much greater accuracy. The amount of stored samples may be lower, but the overall data quality will be improved. Such an approach may find a way for implementation in the other machinery operating in highly variable conditions, like mining equipment or aircrafts.

4.6 Case Study—Validation of Signals from a Wind Farm

The following case study illustrates the performance of the methods presented in this chapter. They have been applied to the data acquired by CMS on a real wind farm. The database contained numerous data acquisition related errors. The research took into account the data from a ten month period. All the turbines have been equipped with a commercial CMS. The data acquisition parameters were: 100 mV/g accelerometers, 25 kHz sampling frequency and 10 s samples length. Each turbine had 6 vibration sensors installed. The validation was performed as presented in the Sect. 4.5 (see Fig. 4.17). All of the algorithm steps were performed for each signal. Even if the test found the signal invalid, all the other tests were completed for test reasons. Table 4.6 presents the test results.

After further investigation on site the system appeared to have serious data acquisition errors which were identified as the insufficient shielding of the cables, the failures of the sensor plugs and the errors of the data acquisition electronic. It is important to note that during the analyzed period only 1 sample a day was stored permanently in the database. The signal features (i.e. trend values of broadband and narrowband features) were calculated and stored in the database every 10 min (e.g. 144 values per day per feature). With the assumption that the analyzed set is a sufficient population sample in the statistical sense it can be concluded that in the worst case (channel No.1), only approximately 1.4% samples (i.e. 500) out of 35,000 are valid and the sensor 1 cannot be a basis for a reliable diagnosis. The second worst sensor was channel No. 5 (generator DE) which experienced only 47.6% of correct samples. This is the number, which might be used for diagnostics, but there is a need to detect and discard all the incorrectly recorded samples. Also, such results create serious doubts about the system reliability and the measurement system should be checked and corrected at the first convenient opportunity.

Table 4.6 Validation algorithm performance on the wind turbine data (reprinted from [5] with permission from Elsevier)

Channel no.	1	2	3	4	5	6
Signals total	249	249	249	250	250	250
Min Energy	205	20	11	8	3	17
PDF fit	102	30	27	23	20	20
N-point	223	34	42	23	24	21
No. of correct signals	4	171	207	220	119	215

References

1. Stander CJ, Heyns PS, Schoombie W (2002) Using vibration monitoring for local fault detection on gears operating under fluctuating load conditions. Mech Syst Sig Process 16(6):1005–1024
2. Yan W, Goebel KF (2003) Sensor validation and fusion for gas turbine vibration monitoring. In: Proceedings of the SPIE conference on system diagnosis and prognosis: security and condition monitoring issues III, vol 5107, Orlando, FA, USA, Apr 2003
3. Alag S, Agogino AM, Morjaria M (2001) A methodology for intelligent sensor measurement, validation, fusion, and fault detection for equipment monitoring and diagnostics. AI EDAM 15(4):307–319
4. Jablonski A, Barszcz T (2013) Validation of vibration measurements for heavy duty machinery diagnostics. Mech Syst Sig Process 38:248–263
5. Jablonski A, Barszcz T, Bielecka M (2011) Automatic validation of vibration signals in wind farm distributed monitoring systems. Measurement 44:1954–1967
6. International Standard Organization (2015) ISO 10816–21:2015. Mechanical vibration—evaluation of machine vibration by measurements on nonrotating parts—part 21: horizontal axis wind turbines with gearbox
7. Zimroz R (2010) Metody Adaptacyjne w Diagnostyce Układów Napędowych Maszyn Górniczych. Oficyna Wydawnicza Politechniki Wrocławskiej, Wrocław (in Polish)
8. Barszcz T, Bielecka M, Bielecki A, Wójcik M (2012) Wind speed modelling using Weierstrass function fitted by a genetic algorithm. J Wind Eng Ind Aerod 109:68–78
9. Barszcz T, Jablonski A (2011) Aspects of automatization of wind farm monitoring on the example of a diagnostic center. In: Proceedings of the 8th international workshop on structural health monitoring (IWSHM) 2:2603–2610, Stanford, CA, USA, 13–15 Sep 2011
10. Jablonski A, Barszcz T (2012) Robust fragmentation of vibration signals for comparative analysis in signal validation, In: Proceedings of 2nd international conference on condition monitoring of machinery in non-stationary operations (CMMNO), Hammamet, Tunisia, 26–28 Mar 2012
11. Antoni J, Bonnardot F, Raad A, El Badaoui M (2004) Cyclostationary modelling of rotating machine vibration signals. Mech Syst Sig Process 18(6):1285–1314
12. Angrisani L, Daponte P, D'Apuzzo M (1999) A method for the automatic detection and measurement of transients—part I: the measurement method. Measur 25(1):19–30
13. Rohatgi VK (1984) Statistical inference. Willey, New York
14. Sazuka N (2007) On the gap between an empirical distribution and an exponential distribution of waiting times for price changes in a financial market. Phys A 376:500–506
15. Magiera R (2002) Models and methods of mathematical statistics. GiS Press, Wroclaw

Chapter 5
Advanced Analysis Methods

5.1 Load Susceptibility Characteristics

5.1.1 Overview

The first method presented in this chapter uses only signal feature data collected by the CMS, after it has been acquired, processed and stored in the system memory. After this process there is no access to the original raw vibration data, as it is not stored due to the extremely large size. Such data is collected on many machines and is typically stored as 10 min averages. The storage is done either in CMS or in SCADA (after it has been transferred there). The method is based on the research for mining equipment by Bartelmus and Zimroz (2009) and later underwent further developments carried out by the author in co-operation with the experts mentioned above [1].

In the simplest approach, fault detection implemented in these systems is performed by simple thresholding on measured values or calculated features (like rms or peak values). However, due to highly nonstationary operation, simple thresholding is not sufficient for early fault detection. The key difficulty is that the stored feature values do not have a single value with a small random factor, as it might be a case for machinery operating in stationary conditions. Instead, the data is non-monotonic and expresses very strong variability. Figure 5.1 presents a set of data covering the period of several months. One can observe that the variability is very high and it is not possible to set a simple threshold value. If such a threshold is set it will lead either to missed faults (when the threshold value is too high) or to many false alarms (when the threshold value is too low). Simple statistical analysis (Fig. 5.2) gives us an interesting insight. In the figure, the probability density functions are given for two features (rms and pp) for two machine states (good and bad). As one can notice the pdfs are non-Gaussian. Moreover, they are multimodal (as the data comes from several different operational states) and they partially overlap (this is the main reason why a single threshold does not yield good results).

© Springer Nature Switzerland AG 2019

T. Barszcz, *Vibration-Based Condition Monitoring of Wind Turbines*, Applied Condition Monitoring 14, https://doi.org/10.1007/978-3-030-05971-2_5

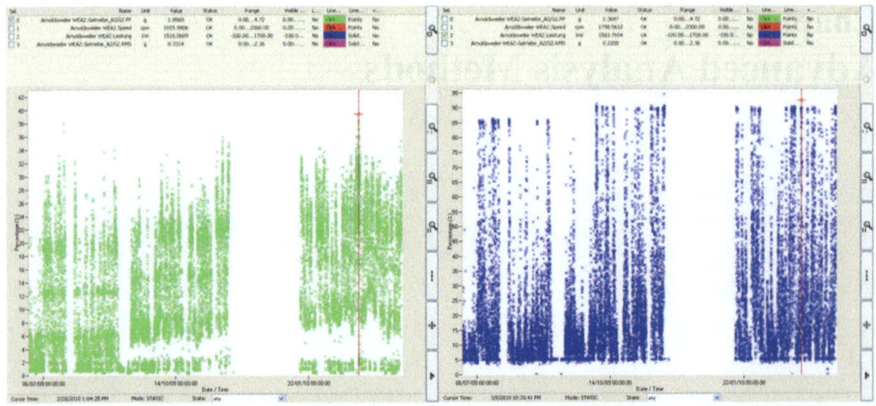

Fig. 5.1 Example of long term data from a wind turbine main bearing: vibration peak to peak (left) and output power (right) (reprinted from [1] with permission from Elsevier)

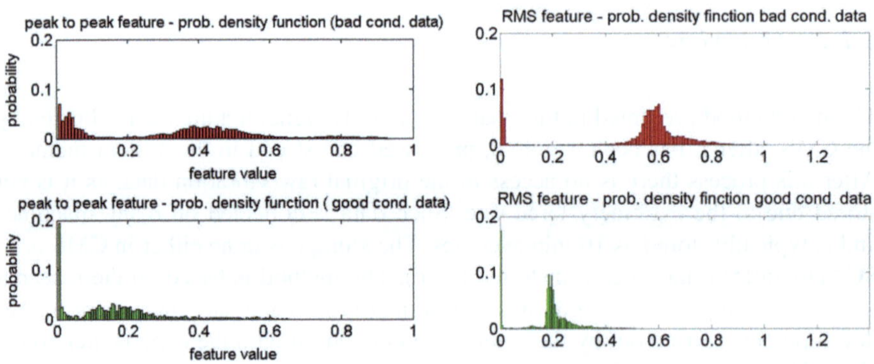

Fig. 5.2 Probability density functions for bad condition (top subplots), good condition (bottom subplots); for left subplots: peak to peak value for main bearing; right subplots: RMS value for generator bearing (reprinted from [1] with permission from Elsevier)

The fundamental assumption of the method was that even when individual features express high variability they depend on an operational point. If we are able to analyze a feature together with another value which can be treated as a measure of the operating point, we can find a dependence between these two. It is possible to express such a dependence in a quantitative way using statistical methods. If the technical condition of a machine changes, the dependence between the signal features and the operational state value will change as well. The implementation of the method has two limitations. It requires to select a value which can be treated as the operating point description. Typically, the best candidate is output power, as it is a parameter which mostly affects the dynamic state. The selection depends heavily on available measurements. The other common candidates can be: rotational speed, wind speed, generator current etc. The other limitation of the method comes from the fact that it needs to collect a dataset to find the dependence between the two values. Therefore, the fault detection requires a longer period of time before the analysis can be performed.

5.1.2 Method Description

Since we are looking for a dependence between the vibration feature and the state variable, one needs to collect data over a sufficiently long period of time. With 10 min averages, such a period should cover at least several months. In the simplest case, the vibration feature can be rms or peak and the state variable—output power or wind speed.

The whole dataset consisting of many 2D vectors should be divided into smaller sets (see Fig. 5.3). To smooth the output values (or to cope with a smaller amount of data) the sets can overlap with each other. Thus, the method applies the moving window approach. An important requirement is that in order to cover the whole feature-state relationship a single set should contain the data from the whole range of state variable (here: load). The authors in [1] proposed to use the length of $N = 1000$ samples. For 10 min values, it is almost exactly equal to 1 week of machine operation.

For each set the concept of feature-load space, proposed by Bartelmus and Zimroz [2] shall be applied. The feature is analysed in the function of power. The simplest dependence is used, i.e. the linear regression. For a single set regression parameters: "a" and "b" are estimated according to the Eq. (5.1):

$$v = a \cdot P + b \tag{5.1}$$

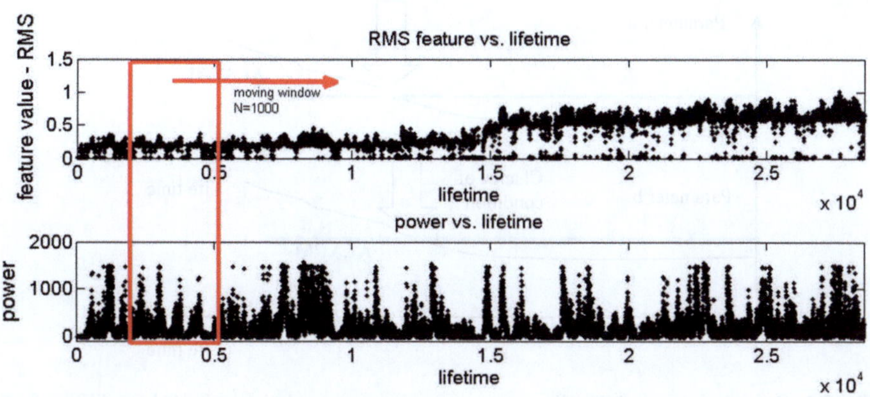

Fig. 5.3 The idea of segmentation RMS data into portions with predetermined length N = 1000 (reprinted from [1] with permission from Elsevier)

where:

v value of a vibration signal feature,
P instantaneous turbine output power.

The output of the method are two value pairs (a, b) per a single set. The result should be independent from the operational state, so it is much easier to interpret by a diagnostic expert. The authors conclude: "When the procedure is finished (i.e. all segments of data have been analysed in the feature-power space), we will obtain a plot describing how parameters of regression were changing during machine lifetime. The change of condition should change the *load susceptibility of machine* (further referred to as LSCh), i.e. distribution of data clouds and the character of *feature-power dependency* should change". The example of trends of both a and b is given in Fig. 5.4. Both LSCh parameters shall respond to the machine fault. The trend of (a, b) pairs is monotonic and it is much simpler to set threshold values individually on a and b values. Despite the variable operating conditions the result is free of random variability and much easier to interpret.

An important advantage of the Load Susceptibility Characteristics method is that it reduces the load dependency of vibration features. Since (a, b) pair describes the dependency, the (a, b) value does not change despite the load change (as long as the sets are properly selected and include sufficient data covering all the turbine load range). If there is fault development, the vibration range might not change, but the relationship will. In general, this change will be non-linear, but LSCh will yield linear approximation of the relationship. The change of LSCh may be even in a or in b parameters. The change in a will mean higher sensitivity of the feature on the load change, whereas the change in b will mean a higher vibration level regardless of the load level. In the previous study about gear damage Bartelmus and Zimroz [2]

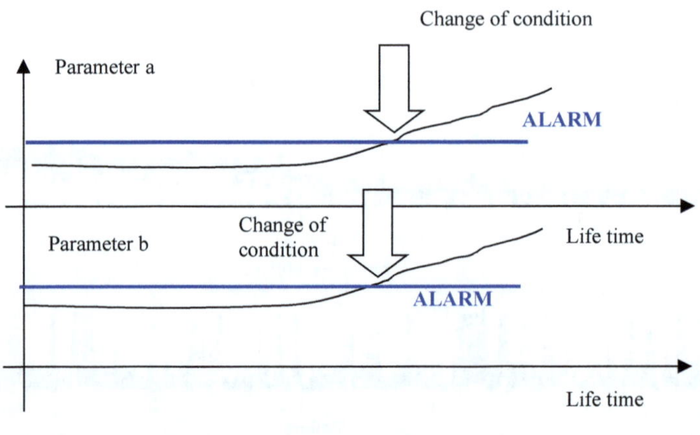

Fig. 5.4 Output of the method—linear regression parameters during recorded machinery fault. Note that the figure is given for illustration purposes and is not based on real data) (reprinted from [1] with permission from Elsevier)

mentioned that the importance of the LSCh came from the observation that in this case there is almost a parallel shift of the LSCh caused by the fault in the diagnosed object.

5.1.3 Case Study—Generator Bearing Fault

The proposed method was demonstrated in the two case studies [1]. The first case was the fault of the generator bearing in a wind turbine. Two trends were selected for analysis:

- RMS of vibration signal taken by the accelerometer on a generator bearing,
- generator output power measured as a process parameter.

Both signals were measured by the CMS and stored as 10 min instantaneous values. The input data is shown in Fig. 5.5. The data axis is expressed in samples and the whole dataset represents the period of c.a. 7 months. Around the time $t = 15{,}000$ a significant change of condition appeared as the generator fault had developed. The rms values are marked with color: green—ok, yellow—fault development, red—alert, black—alarm. The evaluation was provided by the human expert for evaluation purposes.

The described LSCh method was applied to the data. The whole dataset was divided into 28 segments. For each segment the linear regression was performed in the RMS-power space. The data from segments representing good condition and bad condition is presented in Fig. 5.6. The line representing the result of linear regression is presented on each plot. It is clearly visible that both linear dependencies differ in both *a* and *b* parameters, as they are increased for the bad technical state.

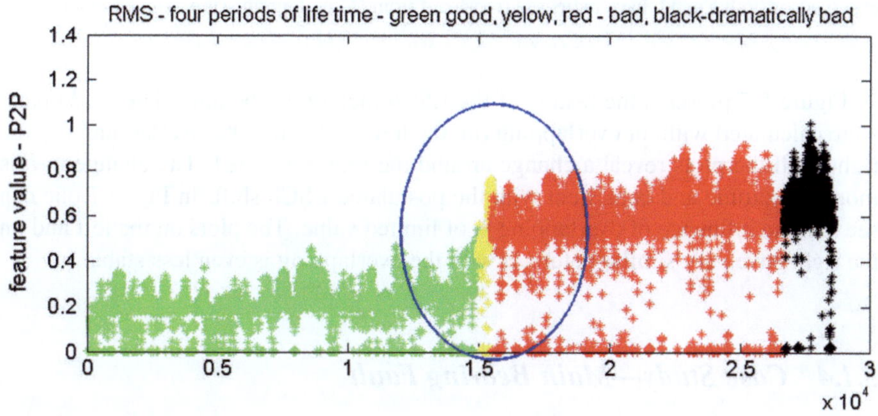

Fig. 5.5 Input data for a generator bearing: RMS long term time series with significant change of condition around $T = 1.5e4$ (reprinted from [1] with permission from Elsevier)

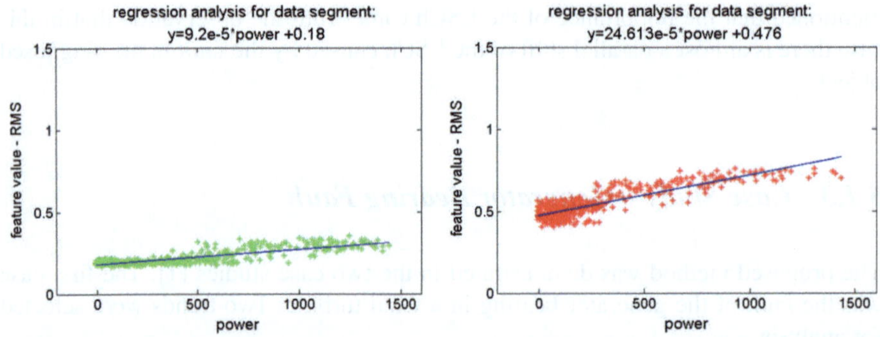

Fig. 5.6 Example of results of regression analysis of two windows: left—for bearing in good condition (data segment no 2) and right—for bearing in bad condition (data segment no 16) (reprinted from [1] with permission from Elsevier)

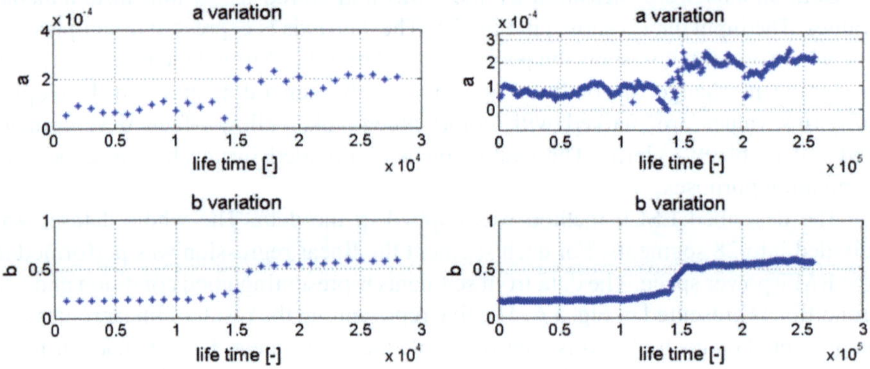

Fig. 5.7 Final data output: variability of coefficients a and b from segment regression analysis. left: without overlapping, right: with overlapping. Note that a clear change (jump) for T = 1.5e4 can be observed, as both a and b have increased (reprinted from [1] with permission from Elsevier)

Figure 5.7 presents the results of the LSCh method to the data. The (a, b) pairs were calculated without overlapping (on the left) and with 90% overlapping (on the right). All the plots reveal a change around the time t = 1.5e4. The change in b is more monotonic and consistent with the postulated LSCh shift. In Fig. 5.7 one can see that the influence of overlapping is of limited value. The plots on the left and on the right have very similar nature, a with the overlapping is even less stable.

5.1.4 Case Study—Main Bearing Fault

The second demonstration case is a faulty main bearing in the wind turbine. The fault was present in the system and was slowly developing. The measured feature was pp of vibration acceleration on the main bearing and the state variable was power,

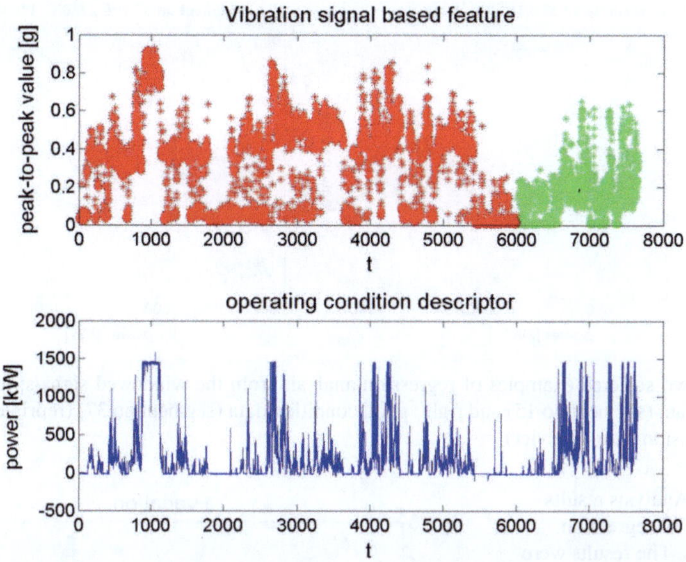

Fig. 5.8 Input data for the main bearing fault: top—vibration PP with a significant change of condition around T = 5500, bottom: generator power (reprinted from [1] with permission from Elsevier)

similarly to the first example. The faulty bearing was replaced and the goal of the test was the detection of this change. The data is presented in Fig. 5.8. A period of short outage is visible around the time t = 5500. The pp values are decreased after this event as well.

The dataset was divided into ca 80 segments covering the whole demonstration period. Figure 5.9 presents the examples of full data from segments representing the bad (left) and the good (right) technical state of the bearing. As presented on the plot, one can observe that the value of a didn't change its value whereas the value of b decreased almost 5 times.

Similarly to the first test, Fig. 5.10 presents the results of the analysis of (a, b) pairs for all the segments with 90% overlap. The parameter a showed large increase just after the replacement, but didn't change the value in the longer term. It is probably caused by the fact that the power after the replacement was very low and the segment didn't contain data which covered a sufficient operating range. The parameter b, on the other hand—showed clear decrease after the replacement of the bearing. Again, these results are coherent with the LSCh properties postulated by Bartelmus and Zimroz [2] shown for the whole data. In this case parameter b decreased significantly after the replacement. Parameter a increased rapidly and it is much more sensitive to damage. However, after the replacement the mean value of a parameter is similar. It is related to a different nature of the degradation process.

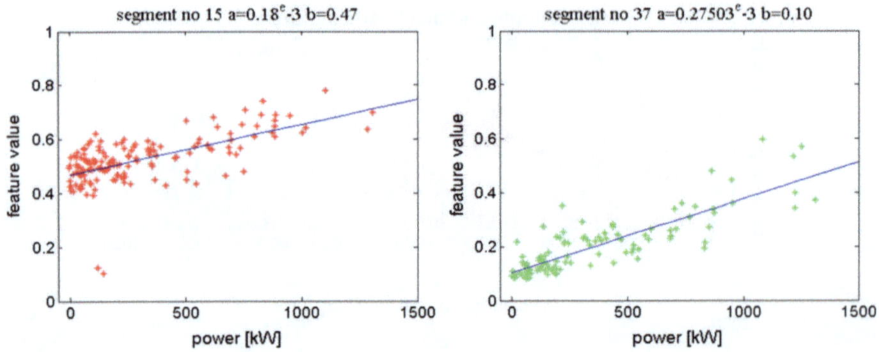

Fig. 5.9 Two selected examples of regression analysis from the windowed signals: left: for bad condition data (segment no 15) and right: good condition data (segment no 37) (reprinted from [1] with permission from Elsevier)

Fig. 5.10 Analysis results for 'a' and 'b' regression coefficients. The results were obtained with overlapping. The change of value in 'b' around $T = 5500$ is clearly visible (reprinted from [1] with permission from Elsevier)

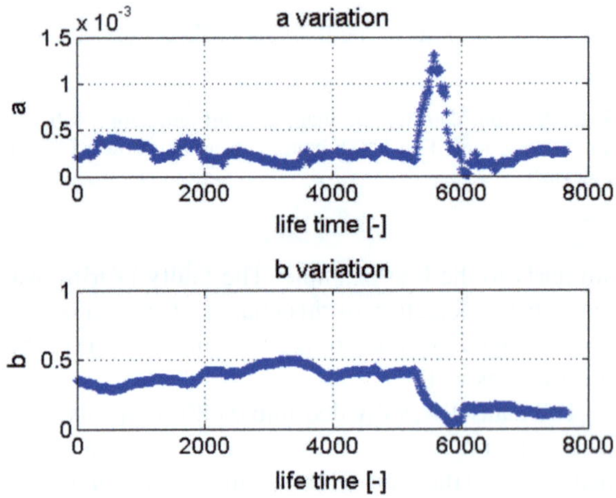

The presented LSCh method which uses the linear regression parameters of relationship between the selected vibration feature and the state variable provides an interesting tool. The method is easy to use and only requires calculated values stored by a vast majority of CMS systems. The presented (a, b) pairs give a much clearer description of the turbine technical state than the vibration related features. These values are load independent and can be used for variable operating conditions.

An important advantage of the method is operating on the data already existing in practically every condition monitoring system. The drawback of the method stems from its statistical nature as it requires to collect enough data points to be able to calculate a and b parameters, thus delaying the detection process. Since the fault development processes are typically slow and take weeks or even months this limitation does not prevent from practical application of the presented method.

5.2 Spectral Kurtosis

5.2.1 Overview

Faults of Rolling Element Bearings (REB) at their early stage generate series of periodic impulses, as presented in the Sect. 1.4. Such an impulse occurs when a rolling element hits a spall on a race. With every passage of a rolling element another impulse is generated and the series of impulses propagates through the machine structure, bearing, shaft and casing. If a vibration sensor is mounted on a casing and the signal is strong enough, one can observe a series of impulses. Apart from REB faults the series of impulses are also generated by other faults. For example gear faults can also generate such signal patterns as it will be presented in the case study later in this chapter. Therefore, detection of peaks in a signal is important for fault detection task.

If the peakedness of a signal is the sought feature there should be an appropriate estimator to detect it. The simplest one is the peak-peak (or the zero-peak), described in the Sect. 2.8. These are broadband estimators implemented probably in every non-basic Condition Monitoring System. Another estimator which is even more sensitive to the peakedness is the kurtosis (see Sect. 2.8). From the statistical point of view, kurtosis K is the 4-th central moment of the time series x (5.2). It is more sensitive to peaks in the signal as it increases the value when the points in the dataset are far outside the average value.

$$K = \frac{\int_{-\infty}^{\infty} (x-\mu)^4 p(x) dx}{\sigma^4} \tag{5.2}$$

where:

μ mean value of x,
$p(x)$ probability density of x,
σ standard deviation of x.

The kurtosis for the Gaussian random series equals 3. There is a definition which subtracts the Gaussian kurtosis value, so it only shows how much the waveform exceeds the Gaussian random noise signal (5.3). This value is often referred to as the Excess Kurtosis.

$$K = \frac{\int_{-\infty}^{\infty} (x-\mu)^4 p(x) dx}{\sigma^4} - 3 \tag{5.3}$$

Kurtosis is an efficient tool for detecting series of peaks as long as the signal does not contain any other strong components. Unfortunately, it is a common case, since many vibration features are at the same time generated by the other machine components, e.g. shafts, gears, blades etc. For example, kurtosis of a signal from several sine signals with small impulse series (as presented in the Sect. 2.2) will be only slightly higher than that for the signal without the impulses. The Narrowband

Envelope Analysis (NEA), as presented in the Sect. 2.4 is a powerful method for detection of impulse series buried in stronger signals. The only requirement is that there exists a frequency band in which the impulsive signal is stronger than the other components. The NEA first performs bandpass filtering of the signal and then calculates the kurtosis of the filter output.

This approach is probably the most efficient way of detecting early stages of REB faults. The drawback of the method is the need to determine the frequency band in which the demodulation (i.e. envelope calculation) is performed. The problem of Optimum Frequency Band (OFB) selection was presented in the Sect. 2.4.2. Spectral Kurtosis (SK) is one of the methods applied to this task and it will be the subject of this chapter. In brief, SK is a computationally intensive method which tests different combinations of filter parameters, which is—for a given filter type—the central frequency and the bandwidth. The returned 2-D map presents a kurtosis value for all the combinations. When a combination with the highest kurtosis value is found, it should be used for signal filtering and next—for envelope calculation.

5.2.2 *Method Description*

Spectral kurtosis is a method which can not only detect non-Gaussian components in a signal, but also determine the frequency of the excited component. The full theoretical background for machine diagnostics using SK can be found in [3]. The examples of application of SK to the diagnostics of bearings and gear monitoring can be found in [4]. Another detailed paper (or rather a mini book) by Antoni [5] presents the subjects in detail with many examples. The SK method is built on interpretation of a vibration signal as a conditionally non stationary process and on its decomposition based on the Wold-Cramer theorem.

Based on this interpretation we consider a system with input $x(t)$ and output of signal $y(t)$. The system is described by a time varying impulse response $h(t, s)$. Then $x(t)$, $h(t, s)$ and $y(t)$ can be presented as:

$$y(t) = \int_{-\infty}^{+\infty} e^{j2\pi ft} H(t, f) dX(f) \tag{5.4}$$

In (5.4) $H(t, f)$ is the time varying transfer function and it can be interpreted as the *complex envelope* of the signal $y(t)$ at frequency f. If H is stationary and both H and X are independent, the process y is conditionally non stationary (CNS). The fourth order spectral cumulant of the process is:

$$C_{4Y}(f) = S_{4Y}(f) - 2S_{2Y}^2(f) \tag{5.5}$$

where $S_{2nY}(t, f)$ is 2nd-order instantaneous moment which is the measure of energy of the complex envelope and is defined as:

$$S_{2nY}(t,f) = \frac{E\{|H(t,f)dX(f)|^{2n}|\omega\}}{df} = |H(t,f)|^{2n} S_{2nX} \tag{5.6}$$

Finally, the SK is defined as the energy normalized cumulant (5.7) which is a measure of the peakedness of the probability density function H:

$$K_Y(f) = \frac{C_{4Y}(f)}{S_{2Y}^2(f)} = \frac{S_{4Y}(f)}{S_{2Y}^2(f)} - 2 \tag{5.7}$$

The practical application of the SK is the Kurtogram proposed by Antoni and Randall [4]. It is a tool with two parameters, namely (a) the center frequency and (b) the bandwidth of the band pass filter. The filter is selected in such a way that it maximizes kurtosis of its output. Kurtogram is the map formed by the STFT-based SK as a function of these two parameters. Thus, the result of the method is a 2-D map. The original kurtogram was based on STFT calculation. It is a very resource-intensive method as all the combinations need to be calculated. In the next step, Antoni [6] proposed to use the filter bank approach. The filter bank calculates filter parameters for a much lower number of variants, but the accuracy of the resulting (freq, band) is on the same level as the original Kurtogram. To differentiate the two approaches, the second map was called the Fast Kurtogram.

Finally, the complete method for fault detection based on Spectral Kurtosis is as follows:

1. Calculate the Fast Kurtogram of the signal (returns CF and BW of the bandpass filter),
2. Filter the signal with obtained filter parameters,
3. Analyze the repetition rate of impulses in the filtered signal (it often requires calculating the envelope of the signal).

5.2.3 Case Study—Planetary Gear Failure Due to a Tooth Fillet Crack

The case study presents the case of a failure in the wind turbine which was presented by the author [7]. The planetary gear experienced a tooth fillet crack on the ring which led to the crack of the gear casing. Figure 5.11 presents the ring gear with the cracked tooth. There was a CMS installed on the turbine, but it did not detect the problem. Figure 5.12 presents the parameters (rotational speed, vibration pp, vibration rms and vibration kurtosis) on the last day before the failure. No increase can be seen on any parameter, except the kurtosis which increased over 10 times, but for the last recorded sample only. It was certainly too late as the controller stopped the turbine.

Since the CMS recorded over 100 vibration parameters, large amount of the data was available for a posteriori analyses. No increase of any characteristic frequency was detected. Fortunately, the CMS also recorded the raw vibration data. Due to its

Fig. 5.11 Picture of the cracked tooth on the ring of the planetary gear. The fully cracked tooth can be seen on the left. The material from the cracked tooth was spread on teeth to the right (reprinted from [7] with permission from Elsevier)

large size it only happened once per day, but altogether there were 60 raw vibration waveforms available for analysis. Out of this amount, 12 waveforms were found to be recorded at the speed of 1500 rpm. Since these signals had higher energy they were selected for further analysis. Figure 5.13 presents the last recorded raw vibration waveform. The signal was acquired 18 h before the failure and does not reveal any symptoms of the oncoming problem.

The spectral analysis did not detect any deviations from a healthy state. The Time Synchronous Averaging (TSA) method was also applied but without success. The main meshing patterns were extracted for different gear stages including the planetary gear. Later, the demodulation of meshing frequencies was also performed. None of the applied methods exhibited fault signatures.

The CMS data contained very long signals covering 40 s sampled at 25 kHz. This was a significant benefit as the planetary gear had very low rotational speed and all its characteristic frequencies were very low as well. Some of the stored raw signals were found to include strong visible peaks (see Fig. 5.14). The period of 1.12 s was repeating between many, though not all, impulses. The 1.12 s period was equal to the ring over roll period and could be linked to the broken tooth development. The ring over roll period is revealed when every planet passing over a point on the ring excites a structural resonance. This should happen in the case of a broken tooth.

The Spectral Kurtosis was used to look for hidden impulses in raw vibration signals. The Fast Kurtogram code implemented in Matlab by J. Antoni was applied to the vibration signals. Figure 5.15 presents the results of the SK algorithm applied to the signal from Fig. 5.14. The resulting plot reveals small but very strong maximum

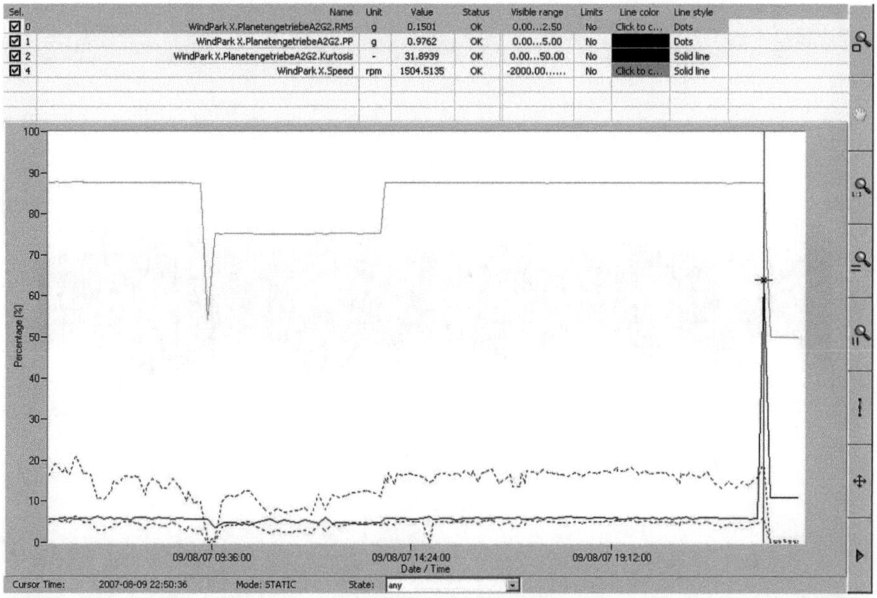

Fig. 5.12 Trend of the key parameters during the last day before the failure. Top line (solid grey) is rotational speed. Note that it is shifted up for clarity. Dotted lines present: rms (lower dotted line) and pp (upper dotted line). The black solid line presents the kurtosis of the signal. Up to the last sample of kurtosis there are no signs of problems. Nothing can be seen even on the last sample on rms nor pp (reprinted from [7] with permission from Elsevier)

where kurtosis raises up to 270.9. This result was obtained for a bandpass filter with central frequency of 11 kHz and bandwidth of only 130 Hz.

When the original signal was filtered with the filter parameters returned by the SK, a signal presented in Fig. 5.16 was obtained. The signal reveals a series of impulses many of which were masked by the other strong signal components. The revealed impulses were found to be periodic and the period of 1.12 s (ring over roll period) was confirmed.

Having detected a series of peaks the same method was applied to all 12 raw vibration signals stored during the discussed period by the CMS. All the signals were acquired at the speed of 1500 rpm, so they could be compared (still not ideally, as the power was not recorded; however, rpm is well related to power). For each signal elevated values of kurtosis in the high frequency band (9–11 kHz) were found. Figure 5.17 presents a trend of maximum kurtosis values in the 9–11 kHz frequency band. The figure shows a rising trend of the kurtosis values rising from 1.5 to 9 to a very high value of 50 after 1.5 months. There are also three data points where the SK reaches values over 200, which is a very high value for kurtosis. Figure 5.18 presents the raw signals filtered with the band pass filter with parameters

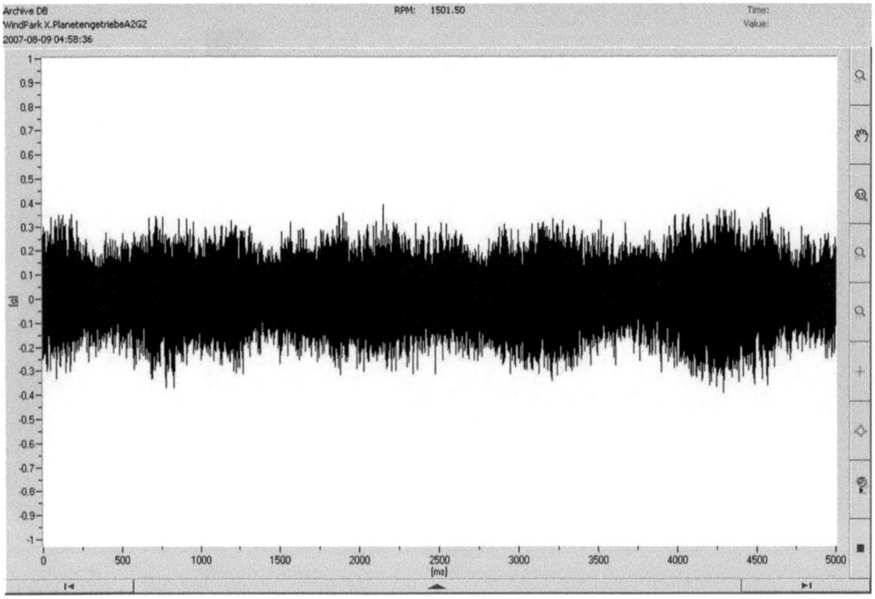

Fig. 5.13 Last stored waveform of operation about 18 h before the failure. No signs of problems are visible (reprinted from [7] with permission from Elsevier)

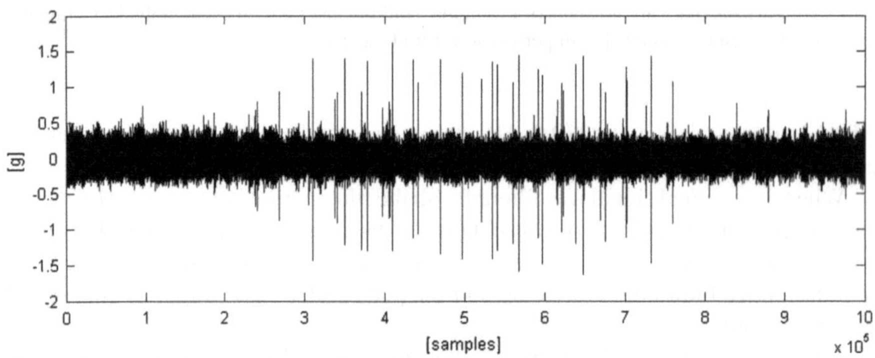

Fig. 5.14 Waveform plot on 28/07. Visual examination of the data revealed existence of high peaks in the signal. Sampling frequency is 25 kHz and total signal length is 40 s (reprinted from [7] with permission from Elsevier)

returned by the SK algorithm. The plots were autoscaled. Thus the noise level may seem different, but in reality it had very similar amplitude. One can see that all the signals reveal peaks. After analysis it was found that peaks are spaced by the ring over-roll period, i.e. 1.12 s.

Automated detection of such a fault cannot be easily performed in the frequency domain, for example with narrowband frequency analysis. It is caused by the fact

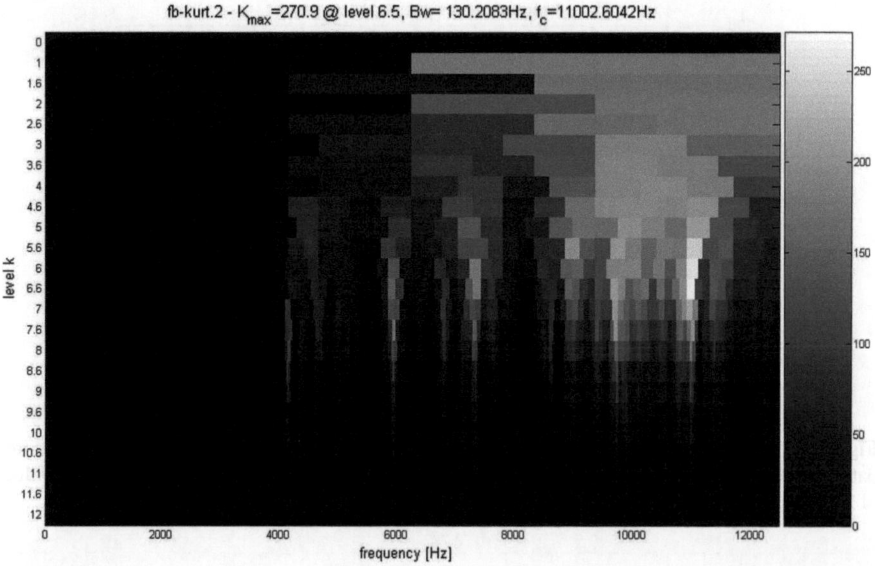

Fig. 5.15 Spectral kurtosis plot obtained from the waveform in Fig. 5.14. Very high maximum (270.9) of spectral kurtosis was detected at 11.0 kHz (reprinted from [7] with permission from Elsevier)

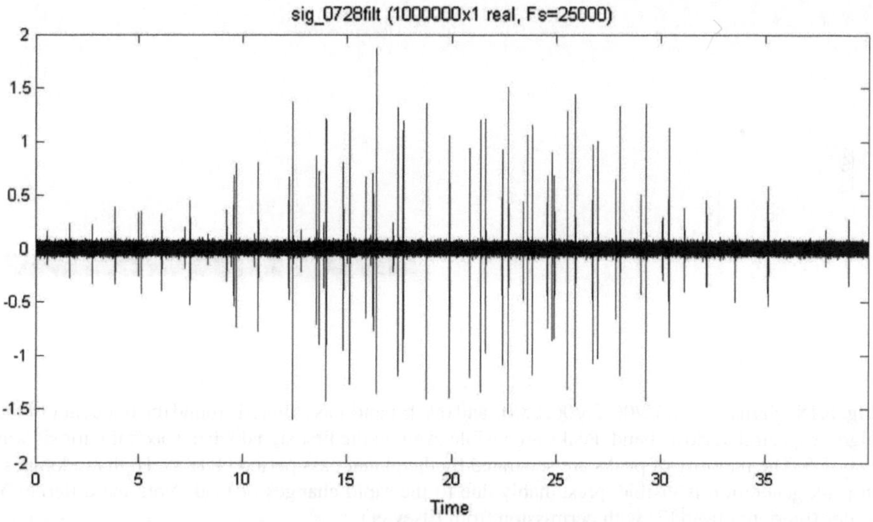

Fig. 5.16 Signal from Fig. 5.14 band-pass filtered around the frequency of the highest spectral kurtosis band (see Fig. 5.15). Several strong peaks are clearly visible. The majority of peaks are separated by the planet pass period (1.12 s). This period exactly reflects passage of the planets over the cracked tooth (reprinted from [7] with permission from Elsevier)

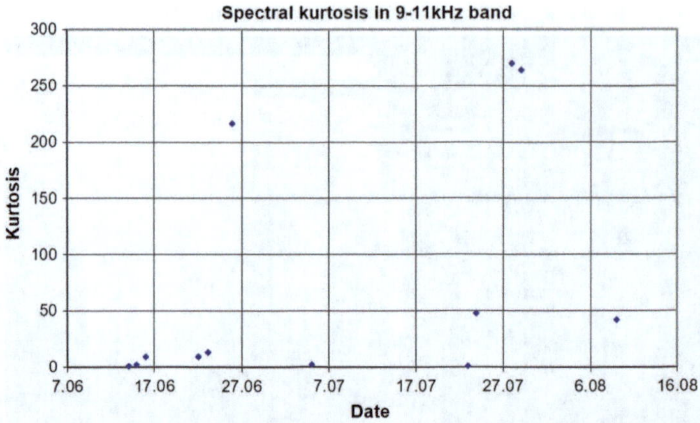

Fig. 5.17 Plot of spectral kurtosis values in the band 9–11 kHz. Despite the low amount of stored data, a trend growing from almost 0–50 is visible. Additionally, three points with very high values of 200+ were stored (reprinted from [7] with permission from Elsevier)

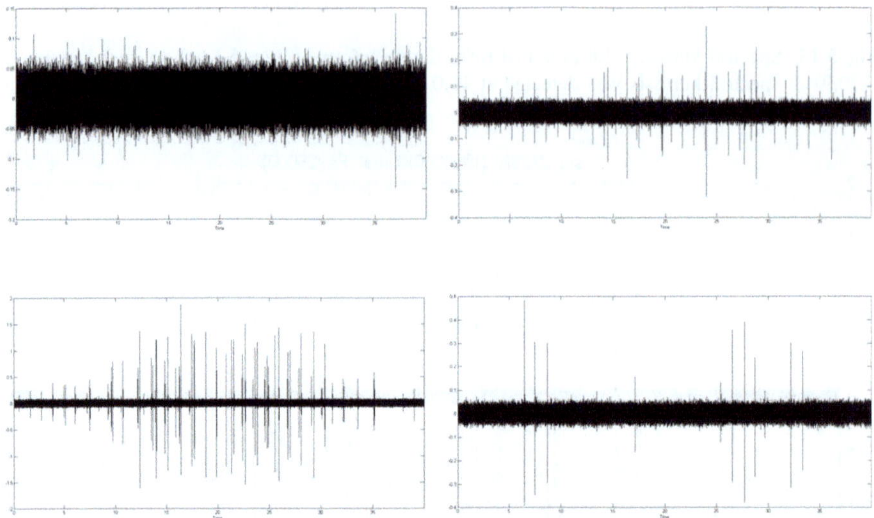

Fig. 5.18 Signals from 16/06, 23/06, 28/07 and 09/08 band-pass filtered around the frequency of the highest spectral kurtosis band. Peaks are visible even on the first signal when spectral kurtosis was only 1.5. The majority of peaks are separated by the planet pass period (1.12 s). High randomness in peak generation is visible, presumably due to the rapid changes of load. Note the different Y scales (reprinted from [7] with permission from Elsevier)

that peaks are of a very short duration, in the range of a few milliseconds and the period between them is about 1 s, so the energy at the sought ring over roll frequency will be negligible. Based on the author's experience, the best method is to perform signal demodulation and look for the signature in the envelope spectrum.

The presented analysis based on the Spectral Kurtosis and the subsequent filtering of the raw vibration signal around the OFB returned by the SK can be a very efficient method. Not only can it detect rolling element bearing faults, but also much more complex and harder to detect cases, like the presented planetary gear tooth fillet crack of the ring gear.

5.3 Protrugram

5.3.1 Overview

In order to extract components which carry information about faults of a rotating machine, amplitude demodulation in some narrow frequency band might be used. In this case the results of amplitude demodulation strongly depend on the selected frequency band or bands. Over the years, many methods have been developed to help to select a proper band. The concept of narrowband amplitude demodulation is based on the same physical characteristics of a vibration signal as classical envelope analysis. However, it enables a more precise selection of demodulated signal bandwidth. As shown by Antoni [3], spectral kurtosis (SK) was shown to help to select such a frequency band. Although SK typically generates valid results it might fail in the presence of a relatively strong or non-Gaussian noise. Another detrimental scenario to SK is relatively a high repetition rate of fault impulses. Such signals are characteristic, for instance, for faulty REB in complex machine trains and highly environmentally affected signals. For the purpose of practical implementation of SK, Antoni introduced the algorithm for fast computation of SK named Fast Kurtogram. Since Fast Kurtogram is vulnerable to the same conditions as SK, it has two minor flaws:

Firstly, for complex signals, Fast Kurtogram frequently shows ambiguous pseudo-optimal confusing frequency bands for demodulation. This is explained by the fact that Fast Kurtogram takes into account a time signal generated from a band-pass filtered signal. In consequence, it shows high values for time signals containing impulses of any kind, regardless of the source. If the highest kurtosis is generated by a component that is not second-order cyclostationary (REB), it is going to point incorrect optimal band.

Secondly, according to Antoni [6], the SK of the process $x(n)$, with an additive noise $b(n)$ is highly sensitive to the noise level:

$$K_{(x+b)}(f) = \frac{K_x(f)}{(1+\rho(f))^2} \tag{5.8}$$

where:

$\rho(f)$ the noise to signal ratio as a function of frequency.

In practice, SK successfully points the optimal band if the sought frequency band is characterized by a low noise, regardless of the high overall signal noise. On one hand, SK is able to detect a fault, even when the overall ρ ratio is high. It is only required that a certain band in which $\rho(f)$ is low enough is present. On the other hand, Eq. (5.7) states that SK decreases rapidly with the noise growth.

In recent years, the concept of selecting optimal demodulation band has evolved in terms of methods capable of coping with signals with higher noise. One of such methods is a so-called Protrugram, also based on the kurtosis value. However, in contrast to the Kurtogram method which calculates kurtosis from bandpass filtered time signals, Protrugram calculates kurtosis of spectral amplitudes of bandpass filtered signals. As shown by Barszcz and Jablonski [11] Protrugram might detect transients with smaller SNR compared with classical, time-based SK.

As illustrated in Courrech and Gaudel [8] resonant frequencies of the housing of rolling element bearings range from 5 to 20 kHz depending on individual structures. Selecting an optimal band is important because in such wide bandwidth, a number of the other components is typically present masking faulty-bearing-induced signal components. The selection of an optimal frequency band might be considered in two interchangeable ways, namely either as the bandwidth start and bandwidth stop or as center frequency and the bandwidth pair. In practice, these parameters are generally not known, because the knowledge of resonant ranges requires investigation of modal parameters during operation. Last but not least, in some cases these impulses are masked by a strong noise excited by external sources.

5.3.2 Method Description

Before describing the definition of the optimum demodulation band we should investigate two parameters of the frequency band, namely the bandwidth and the central frequency. Often, there exists a priori knowledge about machine components for which faults are sought. This information should be taken into account when seeking the band parameters.

Optimal bandwidth selection

In order to select optimal bandwidth, a few factors are considered:

The first factor suggests that the band should be as narrow as possible because in this way the demodulated signal is free from other components. In this scenario the band would include just a single dominant carrier frequency and two sidebands on its both sides clearly visible after amplitude demodulation. However, in practice rolling element bearings induce many resonant frequencies, so this energy is distributed over relatively large frequency range. Furthermore, many algorithms selecting optimal frequency band are based on some kind of a sliding window. Thus, selection of smaller bandwidth dramatically increases time consumption since it requires extra iterations.

The second factor implies bandwidth to be widened. Wide bandwidth provides a large amount of the defect-induced signal energy to be included in the analysis, but at the same time, a larger noise is included as well. Additionally, large bandwidth encapsulates a larger number of bearings to be diagnosed simultaneously which might be a significant aspect for many industrial applications.
The third factor imposes additional constraints on the minimum bandwidth. As stated in [9]: "In bearing diagnostics, it is often desirable to be able to detect up to the third harmonic of the bearing defect frequency in the envelope spectrum. However, this is not always possible as the higher harmonics may have decreased in amplitude to such a stage that they are below the background noise". This applies to a situation in which characteristic bearing frequencies are known a priori. In the same work the authors noted that frequently higher harmonics of characteristic frequencies of bearings are below the background noise.

The presented considerations were analyzed on the data from a test rig with a faulty outer ring of the REB. The BPFO was equal to 144.4 Hz. Figure 5.19 illustrates the influence of the selected bandwidth on the resultant envelope spectra. In this case, the center frequency of dominant resonant frequency was found by modal analysis to be slightly above 4 kHz. In the figure, the bandwidths are going to vary from 200 to 400 Hz while the center frequency is going to be fixed at 4 kHz.

For different bandwidths the visualization of characteristic bearing frequency clearly varies. Generally, it is recommended to view many harmonics of the frequency. However, as illustrated in Fig. 5.19, widening of the bandwidth reduces the amplitude of the dominant fundamental frequency. The plots in Fig. 5.19 lead to the following conclusions:

- the bandwidth must be greater than the sought fault frequency itself,
- if it is too large the resultant envelope spectrum may contain too many distracting components,
- it is recommended to cover 3–5 harmonics.

Optimal center frequency selection

Once the bandwidth is selected, the central frequency for amplitude demodulation needs to be determined. Figure 5.20 illustrates the results of the implementation of different center frequencies with a fixed bandwidth equal to 750 Hz (i.e. slightly more than five sought BPFO equal to 144.4 Hz).

It is visible that as the central frequency approaches the true dominant resonant frequency, the visualization of harmonics of the sought characteristic frequency is much clearer. Thus, for the studied example, the bandwidth around 700 Hz combined with the central frequency around 4000 Hz yields "optimal" results. However, such an approach is not very attractive for commercial systems of monitoring and diagnostics, because it requires a priori knowledge about current structural frequency ranges of individual machine components at a given time, frequently under given operational conditions as well. As a result, practical utilization of amplitude demodulation techniques for REB diagnostics calls for an automatic selection of the optimal band.

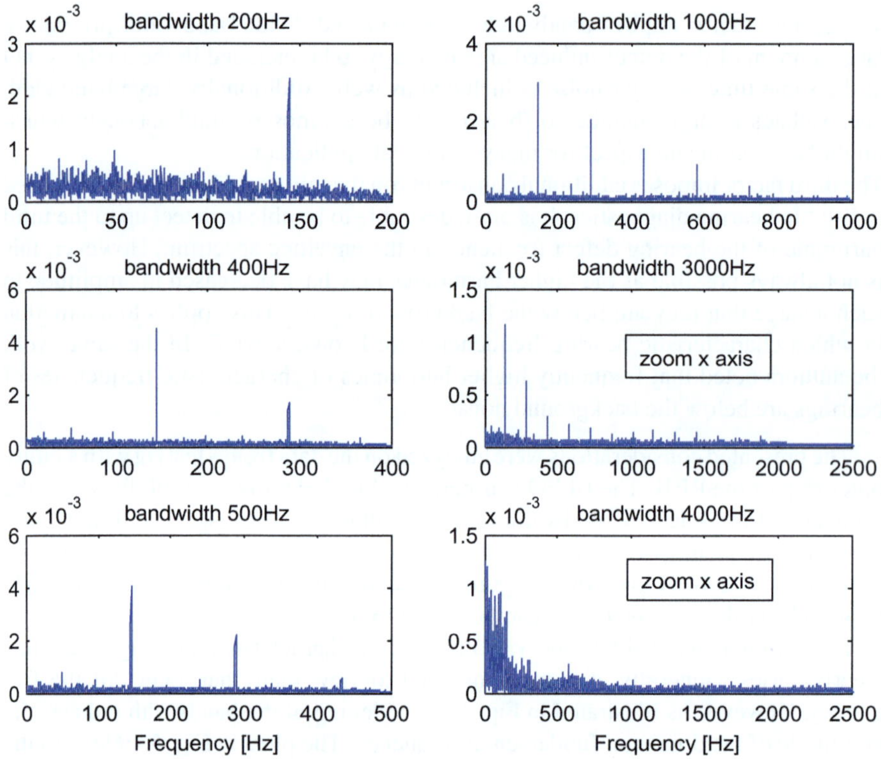

Fig. 5.19 Resultant envelope spectra for 6 different bandwidth values. The central frequency was 4 kHz for all cases

Kurtogram algorithm

In order to explain the algorithm of the Protrugram, it is beneficial to recall the algorithm of the Kurtogram first. Kurtogram presents the matrix of kurtosis value of a filtered version of the original time waveform. The matrix comes from a combination of possible bandwidths and central frequencies which are used as starting and ending frequencies of bandpass filters. Since the number of possible combinations with full resolution is enormous in practical realization of the Kurtogram the combinations are limited according to a particular scheme illustrated by the author of the method in [6]. Because of the nature of the Kurtogram its performance is highly related to the quality of the signal [10]. In order to overcome this potential obstacle, the Protrugram method was proposed by Barszcz and Jablonski [11].

The key description of the Protrugram method is actually quite simple—it is the value of the kurtosis of the envelope spectrum of the investigated signal. As shown before, the Kurtogram uses kurtosis values from the envelope of the filtered signal. The key difference is thus the signal which is used to calculate the kurtosis. In Kurtogram it is the time envelope of the time waveform whereas in Protrugram—it

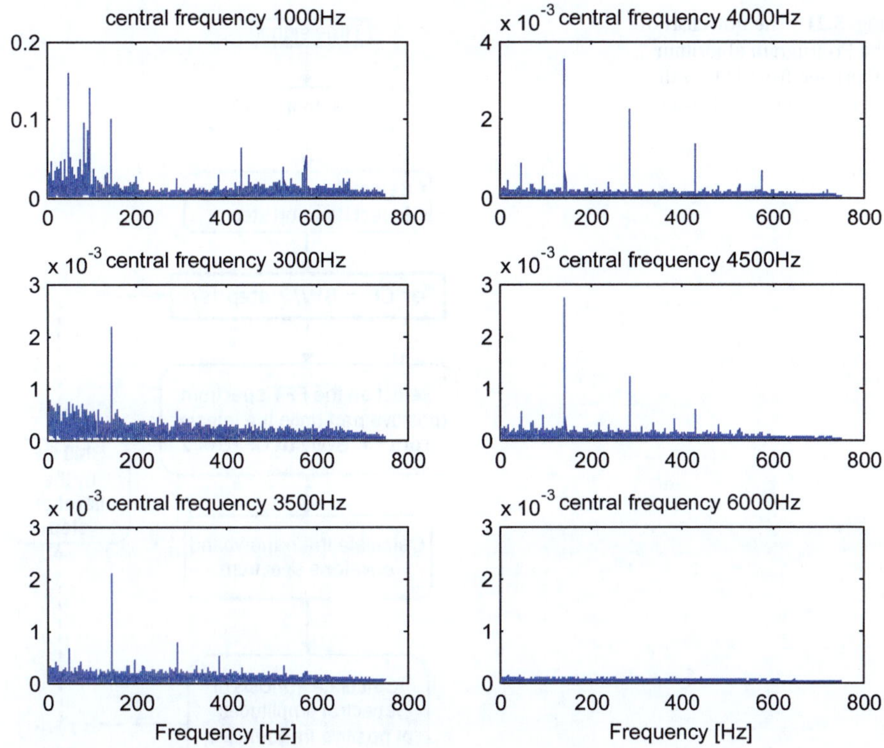

Fig. 5.20 Resultant envelope spectra for 6 different central frequency values. The bandwidth was constant (750 Hz) in all cases

is the envelope spectrum. The key benefit is that the calculation of the spectrum discards any non-periodic components, like e.g. random impacts.

Considering the envelope spectrum from a signal recorded on the bearing housing, along with a growing wear level of the bearing, three categories of the shapes of envelope spectra might be considered:

- envelope spectrum showing just low noise which corresponds to fine REB,
- envelope spectrum with clear peaks and low noise which indicates local REB damage, i.e. incipient stage of the damage,
- envelope spectrum with high noise and no clear peaks which indicates high-level distributed REB damage.

Since the Protrugram operates on the kurtosis of the amplitudes of the envelope spectra, it is mostly designed to show the highest value for local damages. It corresponds to the second point and makes the Protrugram a very efficient tool for detection of REB faults at an early stage.

Let us now investigate the Protrugram calculation in more detail. The method extensively uses kurtosis which was discussed in the Sect. 5.2. The Protrugram

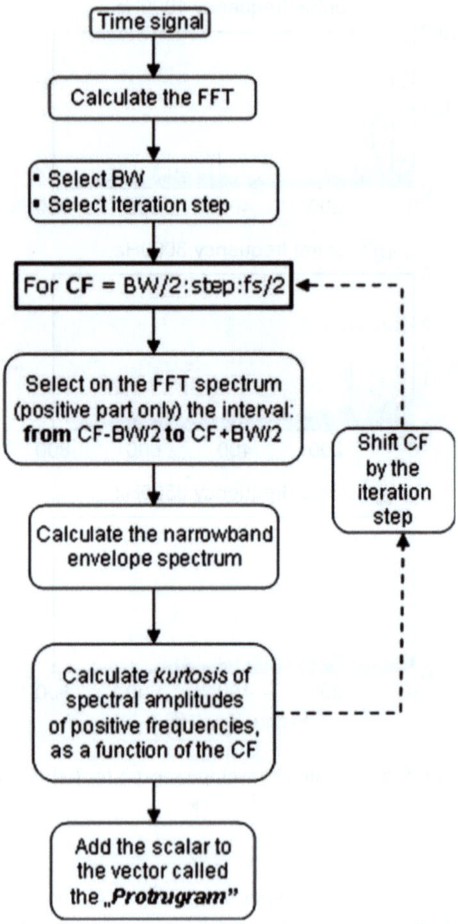

Fig. 5.21 The flowchart of the Protrugram algorithm (reprinted from [11] with permission from Elsevier)

calculates the kurtosis value of a number of calculated narrowband envelope spectra as a function of the center frequency. For each consecutive envelope spectrum, the center frequency (*CF*) is shifted by a known step while the bandwidth (*BW*) is kept constant. The bandwidth could be chosen slightly more than 3× the characteristic frequency of interest. In this way, the input parameters for each step of the Protrugram are obtained. Figure 5.21 illustrates a flowchart of the method along with the flowchart of the Kurtogram. As a result, the Protrugram generates a 2-dimensional kurtosis-frequency characteristics which is displayed as a figure. In case of REB fault the Protrugram will show a hill-like shape around the dominating resonant frequencies. For the advanced identification these zones might serve as an input for narrowband envelope analysis.

When interpreting the Protrugram vectors one can observe that some vectors will continuously increase and then decrease while in the other locations sudden jumps at certain frequencies can be encountered. The first behavior is common for genuine structural resonances when the CF typically gradually approaches the optimal value (enhancing the virtual protrusion of characteristic frequencies above the noise level), and retreats afterwards (protrusion deteriorates). Such a mechanism creates a hill-like shape on the Protrugram vector. The other behavior may be noticed when an individual spectral component is introduced while CF is shifted through the frequency band.

5.3.3 *Case Study—Test Rig REB Fault with Presence of a Single Non-Gaussian Peak*

To present the performance and features of the Protrugram, a signal from a test rig was selected. The test rig was a simple AC motor driving a shaft with a disk, supported on two rolling element bearings. One of these bearings had a seeded fault on the outer ring. The BPFO characteristic frequency of the fault was 144 Hz. The signal was acquired at 24 kHz frequency for a period of 5 s. Figure 5.22 presents the time signal and Fig. 5.23—the linear frequency spectrum of the signal. No resonances were visible in the high frequency range and there were none in the higher bands, either.

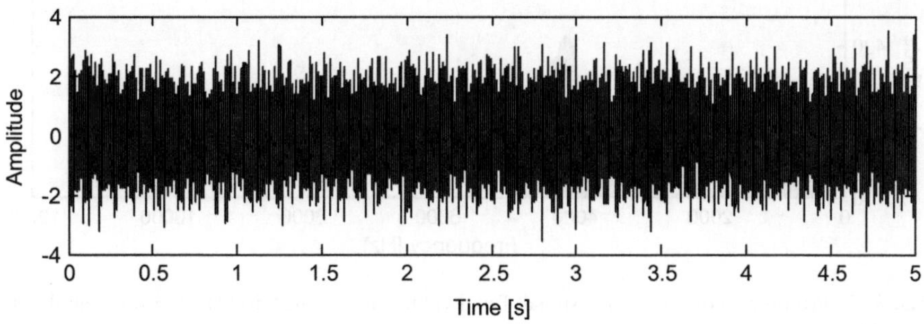

Fig. 5.22 Time view of the test rig signal

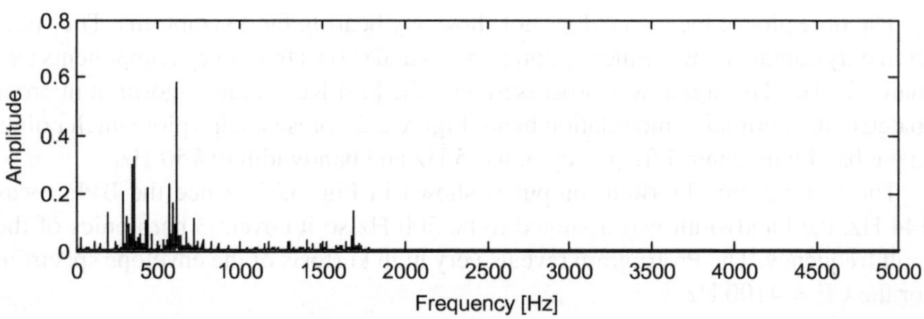

Fig. 5.23 Linear frequency spectrum of the test rig signal

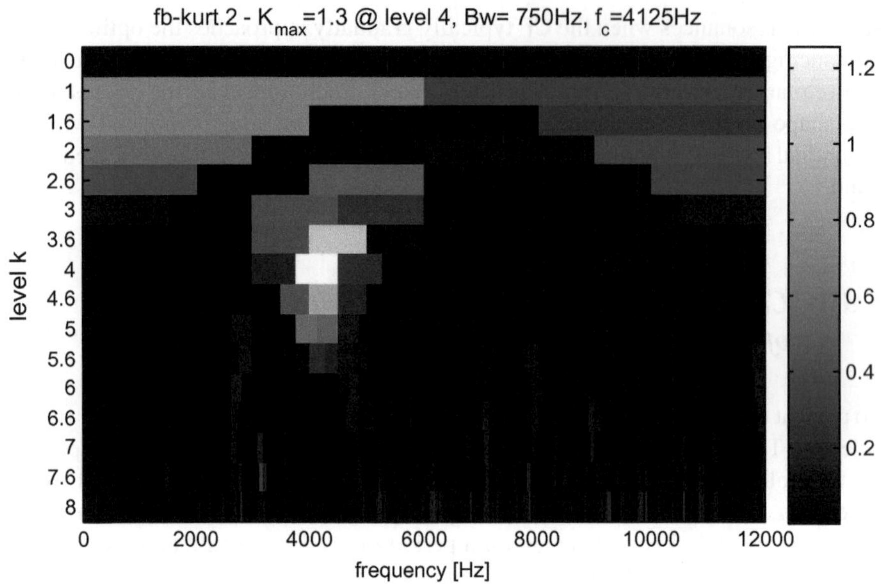

Fig. 5.24 The Fast Kurtogram of the test rig signal

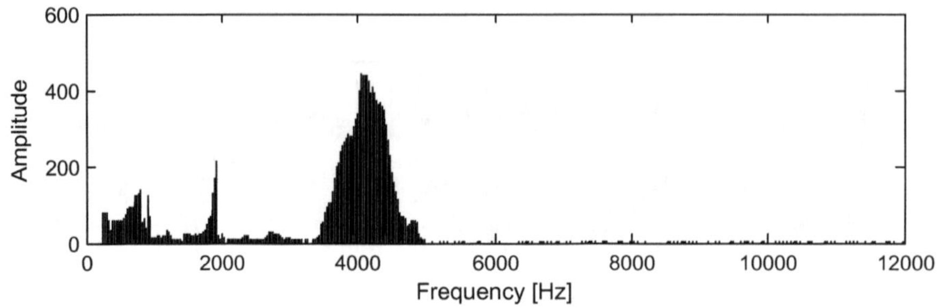

Fig. 5.25 Protrugram of the test rig signal. Bandwidth was assumed 500 Hz. CF was found to be 4100 Hz

The time plot of the signal does not show any bearing fault symptoms. The spectrum only contains low frequency components and shows frequency components less than 1.7 kHz. The signal was processed with the Fast Kurtogram algorithm in order to detect the optimal demodulation band. Figure 5.24 presents this plot which points to the band with central frequency of 4125 Hz and bandwidth of 750 Hz.

The Protrugram algorithm output is shown in Fig. 5.25. Since the BPFO was 144 Hz, the bandwidth was assumed to be 500 Hz so it covers 3 harmonics of the fault frequency. The Protrugram reveals very high kurtosis of the envelope spectrum for the CF = 4100 Hz.

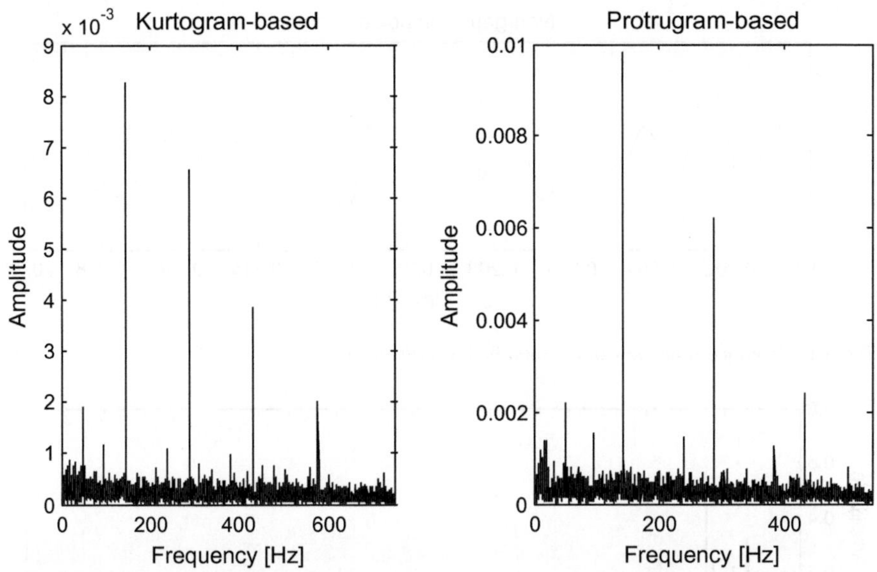

Fig. 5.26 Narrowband envelope spectrum for indicated optimal parameters. Left side—the Kurtogram, right side—the Protrugram

Finally, the parameters returned by both Kurtogram and Protrugram were used for the narrowband envelope demodulation and are shown in Fig. 5.26. Both methods yielded the correct demodulation band and on both envelope spectra a very clear fault pattern can be observed with a strong BPFO line and consecutive harmonics.

Another test was performed on the same signal as in Fig. 5.22, but a single sample of the signal was modified. Figure 5.27 presents a zoomed fragment of the original signal. The sample at time 0.2082 s has a modified value. Such a situation can happen when a short random impact is recorded or an electronic glitch is experienced. The frequency spectrum of the modified signal (Fig. 5.28) cannot be distinguished from the original one. The Fast Kurtogram, however, is very different from the original one (see Fig. 5.29). The proposed demodulation band is FC = 9000 Hz; BW = 6000 Hz and the kurtosis in this band has a huge value of 16,282. The Protrugram shown in Fig. 5.30, on the other hand, was not affected by the signal modification and is practically identical to the original one.

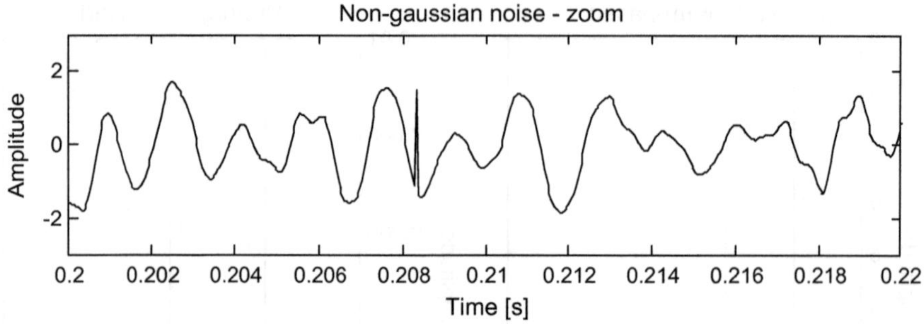

Fig. 5.27 Zoomed time view of the modified test rig signal

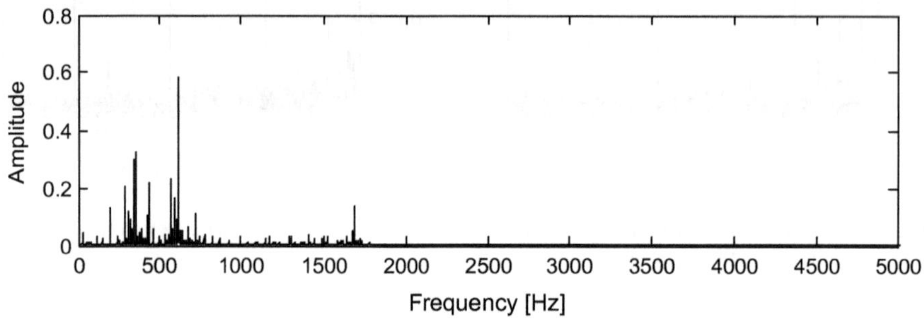

Fig. 5.28 Linear frequency spectrum of the modified test rig signal

Finally, Fig. 5.31 presents two envelope spectra obtained in accordance with the band parameters returned by Kurtogram and Protrugram. The Protrugram was not affected while the Kurtogram version was ruined and contains no useful information. Also the amplitudes of the Kurtogram driven envelope spectrum are much smaller. One could argue that this is not a very realistic case and real impacts (or random spikes) will be of different nature. It is true to some extent only, as the modified sample had, in fact, very small amplitude and is not even visible on the time plot without zooming. Real impacts do happen and have much higher amplitude, though they may have a narrower frequency band.

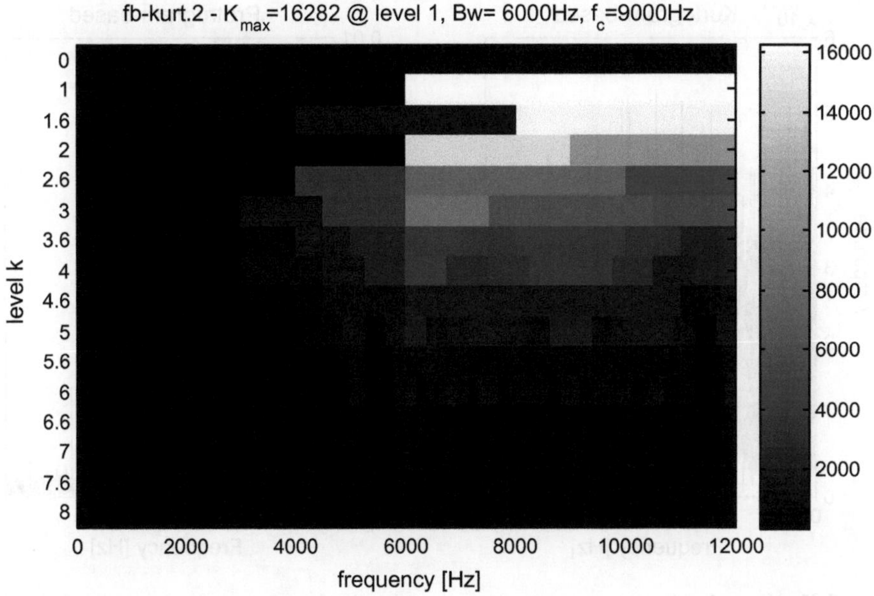

Fig. 5.29 The Fast Kurtogram of the modified test rig signal

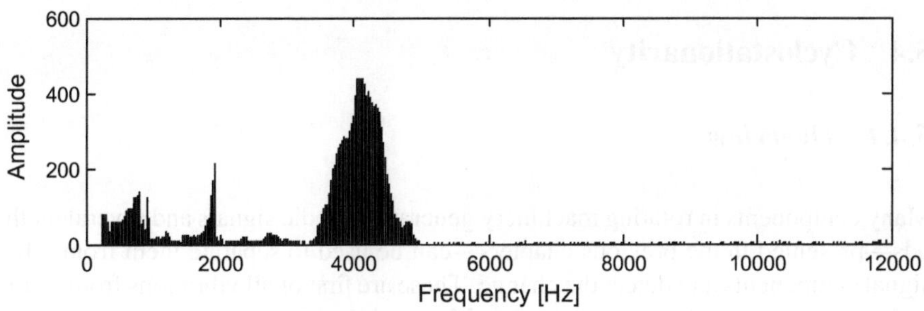

Fig. 5.30 Protrugram of the modified test rig signal. The plot is practically identical to the original one and was not affected by the noise

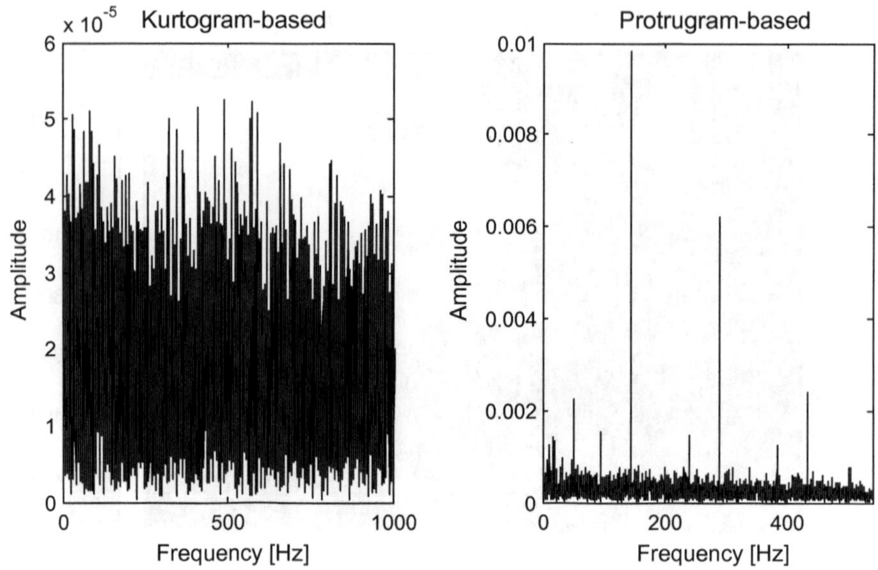

Fig. 5.31 Narrowband envelope spectra for indicated optimal parameters for the modified signal. Left side—the Kurtogram, right side—the Protrugram

5.4 Cyclostationarity

5.4.1 Overview

Many components in rotating machinery generate periodic signals and several methods—presented in the previous chapters—can be used to separate them from other signal components and detect the change. These are first of all vibrations from shafts and gears on constant speed machinery. Many other signals are stationary random, i.e. these signals cannot be predicted, but their statistical parameters, e.g. mean value or standard deviation, remain constant. The best examples are probably fluid related signals found rather in pumps than turbines. There is also another group of random signals where the signal is random, but its statistical parameters vary periodically.

An important insight into these signals can be achieved if one notices that such signals are nonstationary indeed, but some of their properties show periodicities. This approach was given the name of *cyclostationarity* which is a relatively new class of signal analysis methods. Though the method was originally proposed in the 60 s, the original field of research was focused on telecommunication [12]. Its usage in mechanical systems was proposed by Gardner [13]. The interested reader can find a detailed description of these techniques in Gardner et al. [22] and in a very good work by Serpedin et al. [14]. An excellent work based on many examples was presented by

Antoni [5]. This chapter follows the approach and the naming conventions towards cyclostationarity as proposed by J. Antoni.

An important notion for cyclostationary signals is the order of cyclostationarity. In general, a signal whose n-th order statistic is periodic is called n-th order cyclostationary. According to a more formal definition given by Antoni [5]: "A signal is said to exhibit cyclostationarity if there exists a cascade of linear and non-linear transformations that produces a periodic component. It is said to exhibit cyclostationarity at cyclic frequency α if there exists a cascade of linear and non-linear transformations that produces a sine component with frequency α."

In the other words, if there exists a transformation of the order n of the signal so that the transformed signal is periodic, the signal is cyclostationary of the order n. In a majority of cases the transformation can be of the first order (i.e. linear) or the second order. In the first case, cyclostationary signal will have a periodic mean value. Such a signal can be obtained when a periodic signal is added to a noise signal. No non-linear transformation is necessary and linear methods, e.g. frequency analysis can be used for its analysis. The second case is more interesting. In this case the mean value is not periodic and we need to analyze the autocorrelation function and look for its periodicities. In general, the autocorrelation R_{xx} of signal x is given by (5.9).

$$R_{xx}(t,\tau) = E\left[x\left(t-\frac{\tau}{2}\right)x\left(t+\frac{\tau}{2}\right)\right] \tag{5.9}$$

Let us consider a signal in Fig. 5.32.

The signal is result of multiplication of a sine (with 2 s period) by a white noise. The resulting signal is a modulated white noise. Its spectrum will show no signs of periodicity. To verify this assertion Fig. 5.33 presents the PSD of the signal in which no signs of periodicity are present. To see the structure of the signal let us analyze Fig. 5.34 where the spectrogram of this signal is presented. It can be clearly seen that the energy of the signal is changing, but in any given time the signal remains a white noise (spectrum is flat across all the frequencies). The frequency spectrum averages the signal in the duration of the signal and thus it is not capable of detecting the periodicity.

As discussed in the chapter devoted to gears, such a sine modulation will generate sidebands of the carrier frequency which will be spaced from the carrier frequency by the frequency of the modulating sine function. Since the white noise spectrum is uniform the modulation does not influence the spectrum. This type of cyclostationarity is common in rotating machinery especially when wide structural resonances are excited by a periodic force. A good example is periodic impacts generated by a faulty rolling element bearing. They are responsible for excitation of casing resonances when the structural resonances are very broadband and no clear resonance frequency range can be found. From the signal processing point of view, such a situation can be described as a colored noise which is modulated by periodic impulses.

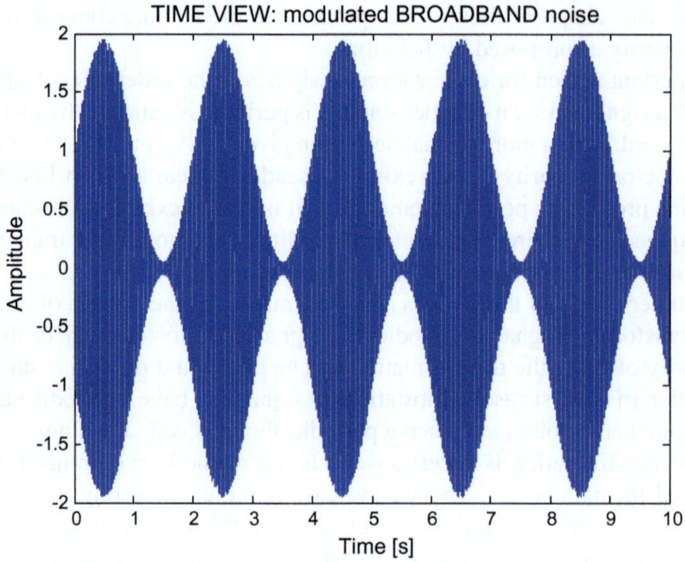

Fig. 5.32 Example of the cyclostationary signal of the second order. The signal is obtained by multiplication of a white noise and a sine function

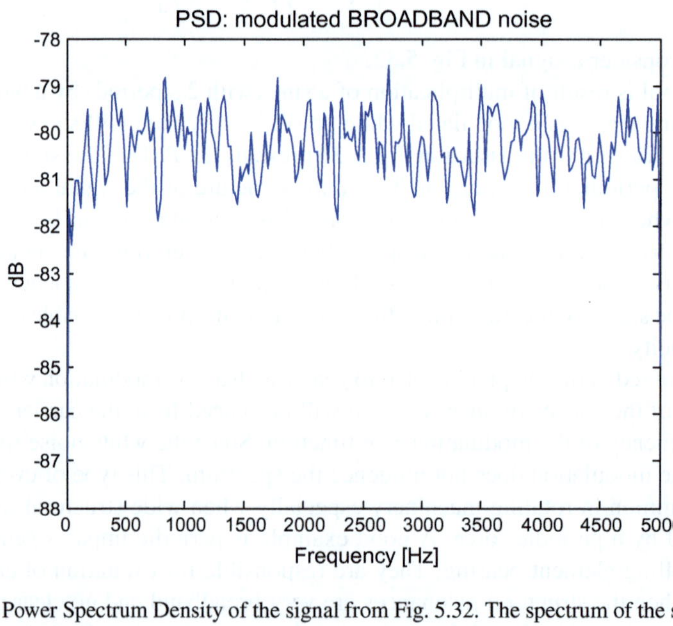

Fig. 5.33 Power Spectrum Density of the signal from Fig. 5.32. The spectrum of the signal shows no sign of periodicity

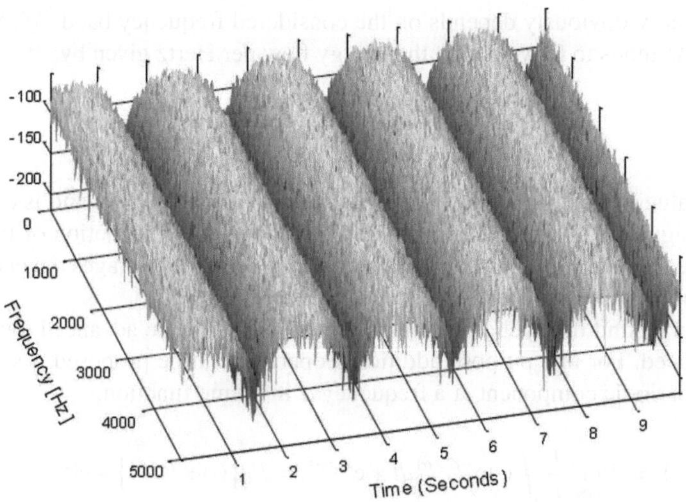

Fig. 5.34 Spectrogram of the signal from Fig. 5.32. Now it is visible that the modulating frequency affects all the frequencies as the carrier is a white noise signal

5.4.2 *Method Description*

A series of analysis tools can be used to find periodicities buried in a signal. This chapter will briefly present several such tools with a rising level of advancement. All the proposed tools start with the notion of a signal power. It is the simplest measure of signal properties and was given in the Sect. 2.8. To present alternative notation, let us first introduce the averaging operator $\mathcal{P}_0$. Let the time averaged power P_x of the signal $x(t)$ be given by:

$$P_x = \mathcal{P}_0\{|x(t)|^2\} = \lim_{T\to\infty} \frac{1}{T} \int_T |x(t)|^2 dt \tag{5.10}$$

The power of the signal is the averaged value of its square. It is not a sufficient measure of any cyclic behavior of the signal as it is simply averaged in time. The common way to analyze periodicities in the signal is the spectrum analysis. Using the averaging approach, the spectrum of the signal can be interpreted as a set of values (or bins) which describes the average flow of the energy in an infinitely narrow frequency band. Let us introduce notation $x_{\Delta f}(t; f)$ which is the signal $x(t)$ filtered through the frequency band with center frequency f of width Δf. For such a signal the average energy flow in that band is given by:

$$P_x(f;\, \Delta f) = \mathcal{P}_0\left\{\left|x_{\Delta f}(t; f)\right|^2\right\} \tag{5.11}$$

The energy obviously depends on the considered frequency band Δf. When the value of Δf tends to 0, we obtain the energy flow per Hertz given by:

$$P_x(f) = \lim_{\Delta f \to 0} \frac{P_x(f;\ \Delta f)}{\Delta f} \tag{5.12}$$

This value is well known as the power spectral density (PSD) and is commonly used for signal analysis. In our example PSD shows the distribution of the energy flow in the frequency domain. As shown in Fig. 5.33, PSD averages the energy flow and will not reveal periodicities of the signal presented above.

In order to find the cyclic phenomena in the signal, more advanced tools should be introduced. For this purpose additional operator can be proposed. Its goal is to extract a periodic component at a frequency α in a time function.

$$\mathcal{P}_{\propto}\{\cdot\} = \lim_{T\to\infty} \frac{1}{T} \int_T (\cdot)\mathrm{e}^{-j2\pi\alpha t} dt * \mathrm{e}^{j2\pi\alpha t} = \mathcal{P}_0\left\{(\cdot)\mathrm{e}^{-j2\pi\alpha t}\right\} * \mathrm{e}^{j2\pi\alpha t} \tag{5.13}$$

The frequency α is referred to as the cyclic frequency of the signal and its reciprocal is the cycle length.

When we want to detect all the cyclic frequencies in the signal, we shall need one more operator which extracts all the cyclic frequencies.

$$\mathcal{P}\{\cdot\} = \sum_{\propto\in A} \mathcal{P}_{\propto}\{\cdot\} \tag{5.14}$$

The operator $\mathcal{P}\{\cdot\}$ will execute the operator $\mathcal{P}_{\propto}\{\cdot\}$ for all the frequencies α from the set A. With help of the newly defined operators, a set of tools can be proposed.

When we want to investigate the energy flow in the signal and to reveal all the periodicities there, the *mean instantaneous power* can be used.

$$P_x(t) = \mathcal{P}\left\{|x(t)|^2\right\} = \sum_{\propto\in A} P_x^{\propto} * \mathrm{e}^{j2\pi\alpha t} \tag{5.15}$$

where:

$$P_x^{\propto} = \mathcal{P}_0\left\{|x(t)|^2 * \mathrm{e}^{-j2\pi\alpha t}\right\} \tag{5.16}$$

The value $P_x^{\propto}$ shall be called the cyclic power. The mean instantaneous power only presents the energy flow without distinction among the spectral content of the signal. It can be useful to reveal hidden periodicity in the signal power, but will not give a deeper insight.

To add the frequency dimension we shall use the *instantaneous power spectrum* (IPS). This tool is defined as:

$$P_x(t, f;\ \Delta f) = \mathcal{P}\left\{\left|x_{\Delta f}(t; f)\right|^2\right\} \tag{5.17}$$

It uses the notation $x_{\Delta f}(t; f)$ which is the filtered version of the signal $x(t)$ with the filter of central frequency f and the bandwidth Δf. The IPS is similar to the spectrogram as it presents the energy flow in both frequency and cyclic frequency domains. Similarly to the spectrogram the IPS is limited by the uncertainty principle:

$$\Delta f \cdot \Delta t \geq \frac{1}{4\pi} \tag{5.18}$$

where the user needs to define the required frequency and time resolutions.

The final step is the *cyclic modulation spectrum* (CMS) which takes one more step from the IPS and reveals the Fourier coefficients of the hidden periodicities. Let us consider Eqs. (5.19) and (5.20):

$$P_x(t, f; \Delta f) = \sum_{\alpha \in A} P_x^{\alpha}(f; \Delta f) * e^{j2\pi\alpha t} \tag{5.19}$$

where:

$$P_x^{\alpha}(f; \Delta f) = \mathcal{P}_0\left\{ \left| x_{\Delta f}(t; f) \right|^2 * e^{-j2\pi\alpha t} \right\} \tag{5.20}$$

Thus, CMS are the Fourier coefficients of the cyclic modulation spectrum of the signal. The resulting picture has two dimensions, namely the spectral frequency f and the cyclic frequency α. The CMS plot is very helpful when the periodicities are hard to spot on the IPS alone. It is also very powerful as it can show whether there are any cyclostationarities in the signal. Moreover, it shows the complete set of cyclic frequencies α for which the energy of the signal is non-zero. Finally, it shows the magnitudes of cyclostationary components at pairs (f, α). The most intuitive interpretation of a (f, α) pair is the cyclostationary component whose carrier frequency f is modulated by the modulating frequency α.

The tools presented above are helpful to introduce the methods of the cyclostationary analysis. The drawback is that neither IPS nor CMS conserve the overall energy of the signal. Thus, another descriptor can be proposed, namely the *Spectral Correlation Density*. The SCD is defined as:

$$SC_x^{\alpha}(f) = \lim_{\Delta f \to 0} \frac{1}{\Delta f} \mathcal{P}_0\left\{ \left(x_{\Delta f}\left(t; f + \frac{\alpha}{2}\right) * x_{\Delta f}^{*}\left(t; f - \frac{\alpha}{2}\right) \right) e^{-j2\pi\alpha t} \right\} \tag{5.21}$$

with usage of another averaging operator $\langle \cdot \rangle$, the equivalent SCD definition can be given as:

$$SC_x^{\alpha}(f) = \lim_{\Delta f \to 0} \frac{1}{\Delta f} \left\langle x_{\Delta f}\left(t; f + \frac{\alpha}{2}\right) x_{\Delta f}^{*}\left(t; f - \frac{\alpha}{2}\right) e^{-j2\pi\alpha t} \right\rangle_t \tag{5.22}$$

The SCD offers a detailed insight into the periodicities in the signal and obeys the energy conservation. Figure 5.35 presents the SCD of the modulated white noise signal from Fig. 5.32. The only non-zero component is the light vertical line. The

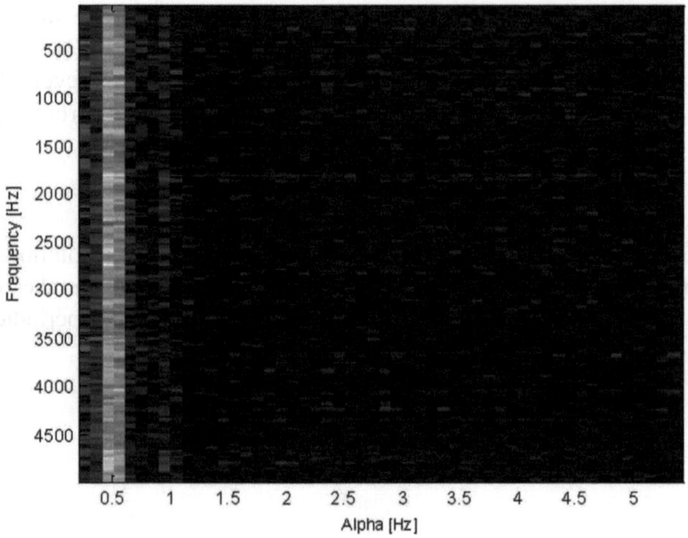

Fig. 5.35 The Spectral Correlation Density of the signal from Fig. 5.32. The strong vertical line shows that the broadband noise ranging from 0 to 5000 Hz is modulated by the cyclic frequency of 0.5 Hz which is exactly the structure of the original signal

line is present for all the carrier frequencies and the cyclic frequency α equals 0.5 Hz. Thus, it presents the broadband noise ranging from 0 to 5000 Hz which is modulated by the cyclic frequency of 0.5 Hz. Therefore, the SCD correctly revealed the structure of the original signal.

The SCD plot can be normalized and the resulting tool is the *Spectral Coherence Density* (SCohD). It is given by the formula:

$$\gamma_x^{\alpha}(f) = \frac{SC_x^{\alpha}(f)}{\sqrt{SC_x^{0}(f + \frac{\alpha}{2})SC_x^{0}(f - \frac{\alpha}{2})}} \tag{5.23}$$

It is interesting to note that the SCohD only detects the pure second order cyclostationarity. It is a very beneficial feature and will be shown in the examples that will follow.

In order to present the advantages of the CS tools for a more complex signal, let us discuss one more example. Figure 5.36 presents the time plot of the signal which was obtained by modulation of two separate colored noises with periodic impulses. Such a signal is close to a real one, for example generated by two faulty rolling element bearings. If bearings are located in different parts of the machine the signal may excite very different structural resonances.

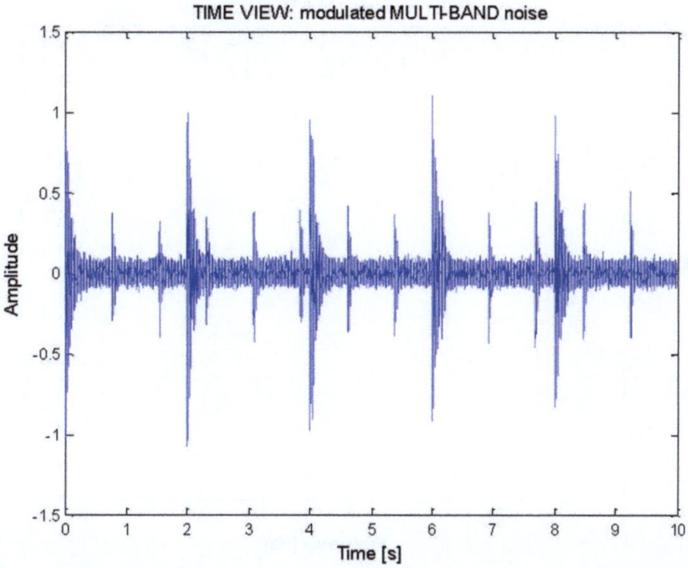

Fig. 5.36 Example of the cyclostationary signal composed of two second order CS signals. The signal is obtained by modulation of two separate colored noises with periodic impulses

Visual examination reveals that it is composed of two series of impulses, one with a period of 2 s and amplitude 1 (zero-peak) and the other one with a period of 0.8 s and amplitude 0.4 (zero-peak). No information about the impulse carrier frequency is visible from the time signal. One can only see that there is a small background noise present.

To get the insight into the frequency structure of the signal, PSD was performed and is presented in Fig. 5.37. It is now visible that there are two frequency bands present, one between 400 and 800 Hz and the second one—between 1500 and 4000 Hz. Unfortunately, all the time information was lost.

To see the internal structure of the signal, let us analyze Fig. 5.38 where the spectrogram of the signal is presented. This plot clearly shows both impulse series. The impulses with the period of 0.8 s excite resonance in the first band (400–800 Hz), while the second series of impulses with the period of 2 s—the resonance in the second band (1500–4000 Hz). The carrier frequency is a color noise (i.e. a filtered white noise).

Figure 5.39 presents the Spectral Correlation Density of the signal. It now clearly visible that both bands contain series of harmonics. The spacing of harmonics is equal to the reciprocal of impulses cycle (1.3 Hz = 1/0.76 s and 0.5 Hz = 1/2 s). The consecutive harmonic lines have the decreasing amplitude. It is a typical frequency spectrum of the series of impulses. Thus, the SCD correctly revealed the impulsive nature of the original signal and detected the carrier frequency bands. Both spectrogram and SCD present the same information, though in complementary domains.

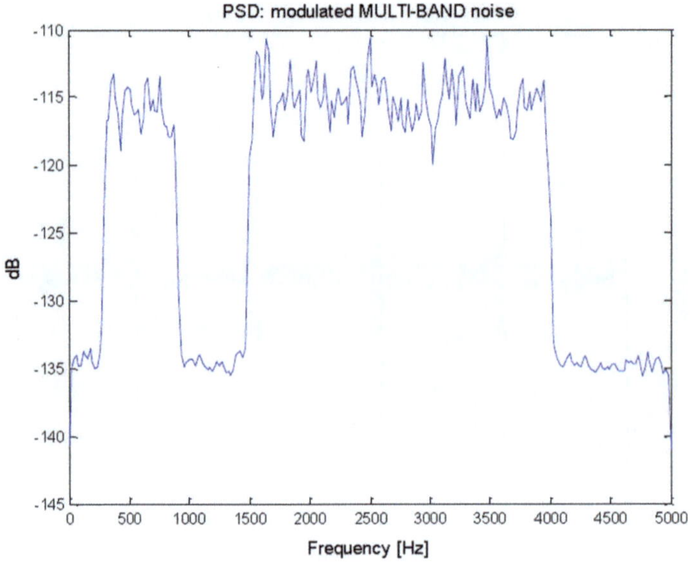

Fig. 5.37 Power Spectrum Density of the signal from Fig. 5.36. There are two color noise bands in the signal, but no information related to time structure. No signs of periodicity are visible

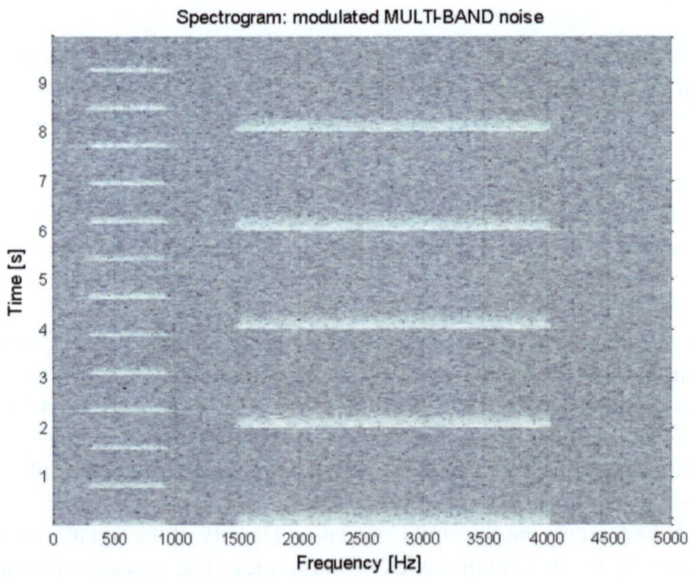

Fig. 5.38 Spectrogram of the signal from Fig. 5.36. Now it is visible that the modulating frequency affects all the frequencies as the carrier is the white noise signal

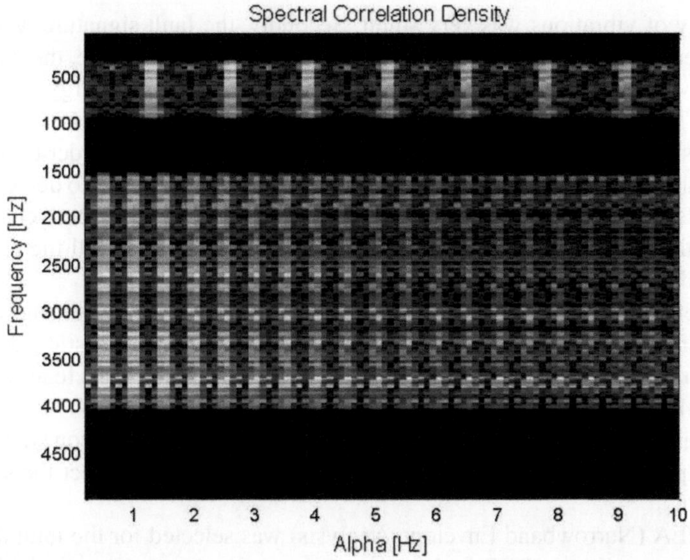

Fig. 5.39 The Spectral Correlation Density of the signal from Fig. 5.36. Both bands 400–800 Hz and 1500–4000 Hz contain series of harmonics, thus reveals the impulsive nature of the modulation and correctly reflects the original signal

What benefit does the SCD bring then? The major difference is the information about the spectral content of the signal. One can notice that it is visible on both figures. It is true, but only because the signal is still simple. For clarity reasons, no other signal components were added. In a real signal there are several harmonic low frequency components generated by shafts and gears together with numerous harmonics. Moreover, the noise level is much higher and makes visual identification of components on the spectrogram much harder, if not impossible.

5.4.3 Case Study—Application of Spectral Coherence for Detection of the REB Fault

Apart from the simple examples presented in the previous chapter, let us consider a real life example of a complex rolling element bearing fault, published by Urbanek et al. [15]. The fault on the inner race of the bearing was found during the endoscope test and was so severe that the machine needed to be stopped in order to exchange the bearing. The investigated wind turbine had a common design of a planetary gear and a two stage parallel gear. The fault was located on the bearing on the shaft connecting the planetary gear and the parallel gear (called the slow shaft of the gearbox). There were two factors which made the fault detection very hard. Firstly, due to the slow rotational speed (14 times slower than that of the generator shaft), the energy of vibrations was very small. Secondly, the fault signature was masked by another bearing fault—located on the generator shaft. Therefore, the turbine had two faulty bearings—the one on the generator shaft had high energy of generated impacts. This type of faults is quite frequent and easy to detect. The other one, on the slow shaft, created a bigger problem to the turbine and was harder to detect.

The turbine was equipped in a CMS system, but it was not able to detect the slow shaft REB fault. Figures 5.40 and 5.41 present the trends of—respectively—PP and RMS of the vibration acceleration from the sensor on the gear. All the plots cover the period of two years.

Neither of the plots shows the increase in the signal. Moreover, the CMS did monitor the narrowband feature configured to track the characteristic frequency of the inner race (BPFI) of the REB on the slow shaft. The trend of this feature is shown in Fig. 5.42. It doesn't show any increase, either.

The benefit of the installed CMS was that it stored the raw vibration signals which could be post-processed to search for a method being able to detect the slow shaft REB fault.

The NEA (Narrowband Envelope Analysis) was selected for the fault detection, as it is very sensitive to REB faults, even at very early stages. This method requires the frequency band which is used for the signal demodulation. First, the spectra were investigated to check visually whether there were any resonances in which the signal could be demodulated. Figure 5.43 presents comparison of the PSD for spectrum from the time shortly after the CMS commissioning and 20 months later during the time period right after the endoscopic investigation. There was a small increase in

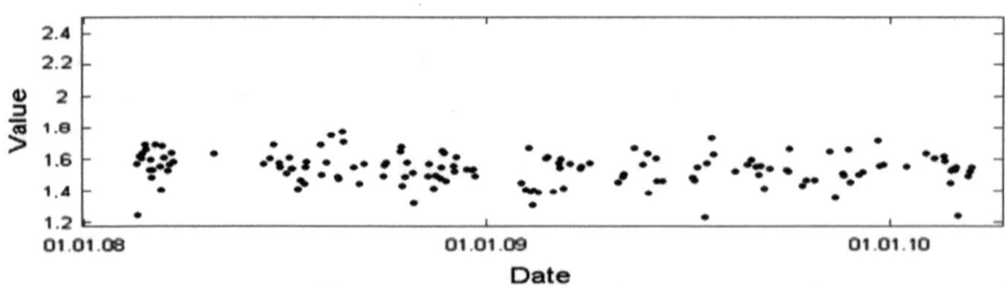

Fig. 5.40 The trend of the PP feature of vibration acceleration

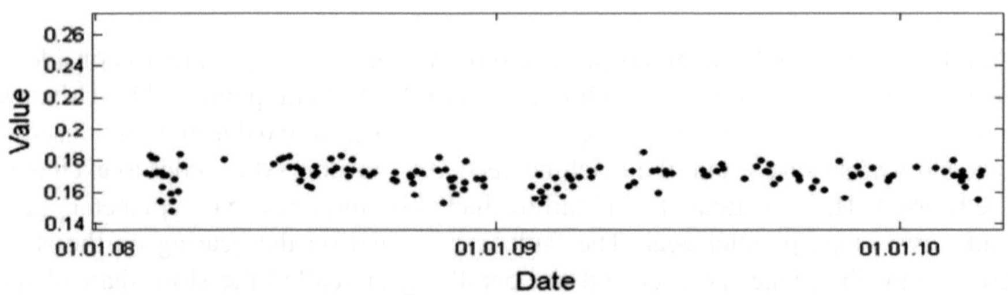

Fig. 5.41 The trend of the RMS feature of vibration acceleration

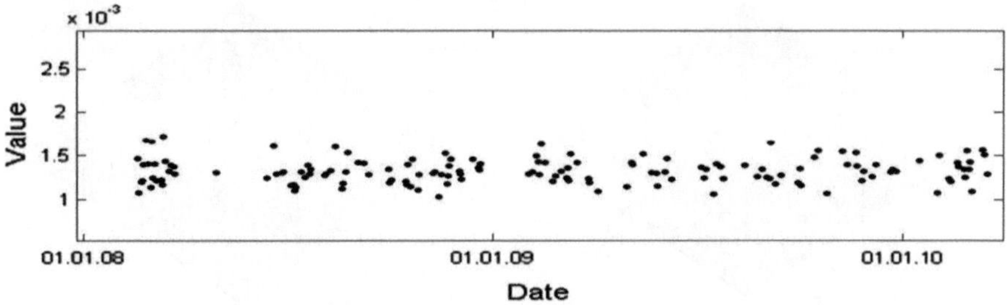

Fig. 5.42 The trend of the BPFI feature of vibration acceleration

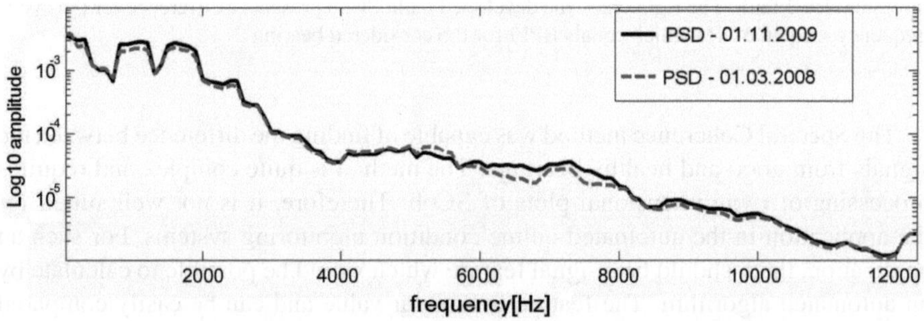

Fig. 5.43 Comparison of PSD spectra shortly after commissioning of the CMS and from the period when the slow shaft REB fault was found. There was a slight increase in the band 6–8 kHz, but was found related to the generator shaft REB fault

the band between 6 and 8 kHz. Based on the NEA in this band, it was found that the increase was caused by the development of the generator bearing fault mentioned earlier.

The next method was the Fast Kurtogram which was applied to the investigated signals. The kurtograms for raw signals were very similar. Apparently, there were no new modulations in the signal caused by the REB fault compared with the rich frequency content of the original signals.

The next method was the Spectral Coherence (SCoh) which was presented in the previous chapter. It is a tool suited for the detection of the second order cyclostationarities. Figure 5.44 gives the comparison of the SCoh for two signals presented in Fig. 5.43.

For better clarity, Fig. 5.44 was zoomed to show only the fragment of the SCoh plot where the difference is visible. The plot on the left was taken on the turbine just after commissioning of the CMS and the one on the right—when the endoscopic investigation found the slow shaft REB fault. Comparison of both plots reveals the difference for the cyclic frequency α equal to 31 Hz. This frequency was found to match the BPFI, i.e. the characteristic frequency of the inner ring of the bearing in question.

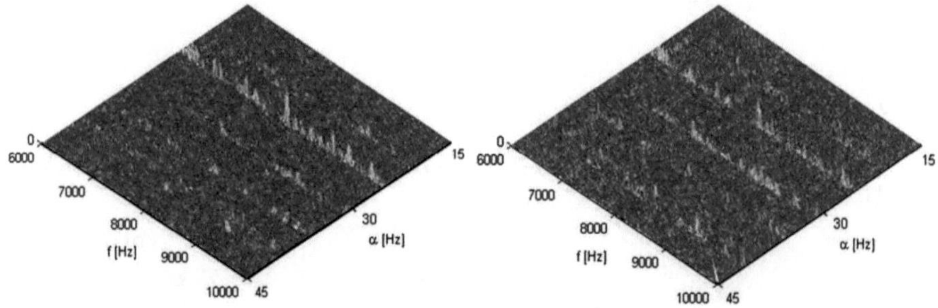

Fig. 5.44 Comparison of Spectral Coherence for signals from Fig. 5.43. The left one indicates no slow shaft REB fault. The right one—the developed fault. SCoh presents a difference for the cyclic frequency α equal 31 Hz which equals BPFI for the considered bearing

The Spectral Coherence method was capable of finding the difference between the signals from good and healthy bearings. The method is quite complex and requires processing of two-dimensional plots of SCoh. Therefore, it is not well suited for the application in the automated online condition monitoring systems. For such an application, there should be a signal feature which would be possible to calculate by an automated algorithm. The feature is a scalar value and can be easily compared versus a threshold level. After the investigation of SCoh plots it was found that the best band for the demodulation was between 8900 and 9500 Hz. After the demodulation and calculation of the envelope spectrum from the resulting signal, the energy of the characteristic frequency (here 31 Hz) should be taken as the sought feature. Figure 5.45 presents the trend of this feature calculated for the raw vibration signals stored by the CMS.

The results have large variability, as even in the first period the values change from 60 to 200. Such a behavior is caused by varying operational conditions and is typical for wind turbine vibration features. The variability starts to grow around 04.2009 when the feature level rises over 250 for the first time. It is hard to decide whether this moment is the beginning of the REB fault, but since the feature is closely related to the BPFI it is a probable hypothesis. Adding more operational states could

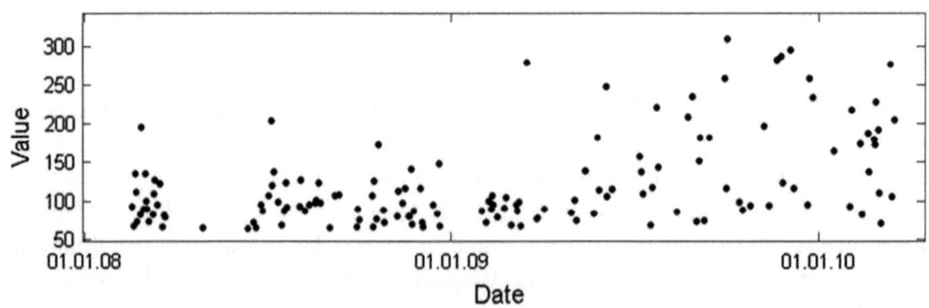

Fig. 5.45 Trend of the feature proposed for the detection of the slow shaft REB fault

further increase the accuracy of the method, but requires more data stored by a CMS including operational parameters. Unfortunately, it was not available for the investigated data.

5.5 Modulation Intensity Distribution

5.5.1 Overview

As presented in the previous chapters, many machinery faults induce modulations into the vibration signals and so, if we want to detect a machinery failure—the best method is to find modulations hidden in the acquired vibration signals, as shown by Nandi [23] and Antoni [5]. In the engineering practice a very efficient method is the Narrowband Envelope Analysis which was presented in the Sect. 2.4. When the optimum band for the signal demodulation is known, bearing and gear faults can be detected with great accuracy. This chapter will present yet another very efficient method for modulation detection. It can yield good results for both signals e.g. when the carrier frequency is a harmonic signal as well as when it is a noise. In the latter case, we obtain a cyclostationary signal and the methods presented in the previous chapter will be applied for its analysis.

In the search for modulations up to this point the sought information was the cyclic (modulating) frequency, as it reflects the impulse repetition rate for the case of REB faults. On the other hand, the information about the carrier frequency is also worth considering. It depends on the machine structure and determines the transfer path between the source of cyclic frequency and the sensor. For different faults and different sensors we can expect that the transfer paths will be different. On the one hand, information about the frequencies allows for proper demodulation, but on the other hand, it can be used for monitoring of structural properties of a machine.

The main benefit of the Modulation Intensity Distribution (MID) method is the two dimensional map revealing dependencies between cyclic frequencies and their carrier frequencies where both pieces of information—about the carrier frequencies and the modulating frequencies—are presented at the same time on two axes. The information contained on the map presents all the modulation contained in the signal. Thus, the name of the method was coined to reflect the distribution of modulations. The original paper proposing the method was published by Urbanek et al. [16].

5.5.2 Method Description

On the frequency spectrum, the modulation is represented by a carrier frequency component with a number of symmetrical sidebands. The sidebands are spaced between the carrier frequency by integer multiplies of the cyclic frequency α. The key idea for the method is that these frequencies are present in the spectrum at the same time. Therefore, the method should investigate for presence of 3 spectral lines with spacing equal to the cyclic frequency α.

Sideband filter

The first part of the method is the application of the "sideband filter" which is a filter ensemble able at the same time to extract information from three frequency bands, as presented in Fig. 5.46. The sideband filter has three parameters, namely:

- f—center frequency of the middle band of the filter,
- α—spacing between the middle band and the sidebands,
- Δf—filter bandwidth (for a single band).

After the filtration (with the ideal filter), we obtain the carrier frequency band and the pair of first sidebands only. It can be presented as:

$$x_i = x_{\Delta f}(t; f - i\alpha), \quad \text{for } i = \{-1, 0, 1\} \tag{5.24}$$

where $x_{\Delta f}(t; f - i\alpha)$ stands for the filtered version of $x(t)$ in the narrow frequency band $[f - \Delta f/2; f + \Delta f/2]$. Since all the other bands will be discarded, the signal to noise ratio will be greatly increased.

In the general case, there is no "a priori" information about the location of the carrier frequencies f nor modulating frequencies α. Therefore, the full $[f, \alpha]$ space should be searched with the "exhaustive search" algorithm, i.e. trying all the combinations of f and α values. The possible algorithm for such a search is presented in Fig. 5.47. The sideband filter for the $[f, \alpha]$ space is calculated with the f and α sweeping across their scopes and the Δf being the parameter.

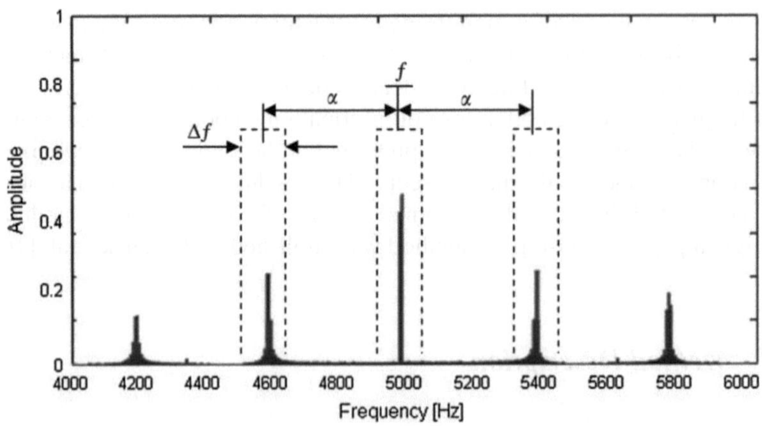

Fig. 5.46 Example of the frequency gain of the proposed "sideband filter" (reprinted from [16] with permission from Elsevier)

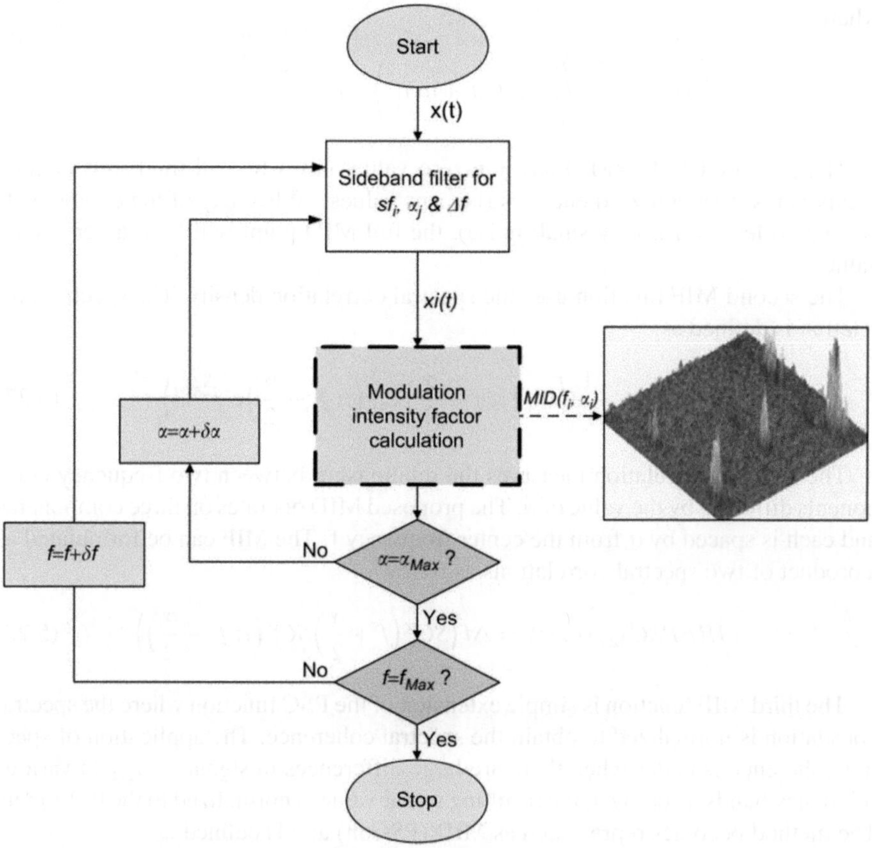

Fig. 5.47 Proposed algorithm for calculation of the modulation intensity distribution (MID) map (reprinted from [16] with permission from Elsevier)

Modulation intensity factor

The second part of the method is the function which shall transform the result of the filter ensemble into a single scalar value. The name of the function is the *modulation intensity factor* (*MIF*) and represents one point on the 2-D MID map. Various functions can be used as the modulation intensity factor and this choice will significantly affect features of the method. In order to keep track of the MIF function applied, the name of the MID method is accompanied by the symbol of the MIF function in the form MID(*method*). A few possible functions of its kind were proposed in [16].

The first function for the MIF is the multiplication of energies in all three bands. The RMS value is good representation of the signal energy and thus MID(RMS) can be defined as:

$$MID(RMS)^{x_i}_{\Delta f} = \sigma_{\Delta f}(f - \alpha) \cdot \sigma_{\Delta f}(f) \cdot \sigma_{\Delta f}(f + \alpha) \tag{5.25}$$

where:

$$\sigma_{\Delta f}^{2}(f+i\alpha)=\left\langle \left| x_{\Delta f}^{*}(t,f+i\alpha)\right|^{2}\right\rangle_{t} \quad for\, i=\{-1,0,+1\} \tag{5.26}$$

The proposed MID(RMS) has non-zero value only when all the bands contain components with non-zero energies (i.e. rms values). When any of the components has a zero level (or a very small noise), the full MID point will have a very small value.

The second MIF function uses the spectral correlation density. The spectral correlation is defined as:

$$SC_{x}^{\alpha}(f)=\lim_{\Delta f\to 0}\frac{1}{\Delta f}\left\langle x_{\Delta f}\left(t;f+\frac{\alpha}{2}\right)x_{\Delta f}^{*}\left(t;f-\frac{\alpha}{2}\right)\mathrm{e}^{-j2\pi\alpha t}\right\rangle_{t} \tag{5.27}$$

The spectral correlation measures the relationship between two frequency components differing by the value of α. The proposed MID operates on three components and each is spaced by α from the center frequency f. The MIF can be formulated as a product of two spectral correlations as follows:

$$MID(PSC)_{\Delta f}(f;\alpha)=\Delta f\left\langle SC_{x}^{\alpha}\left(f+\frac{\alpha}{2}\right)SC_{x}^{\alpha}\left(t;f-\frac{\alpha}{2}\right)\right\rangle \tag{5.28}$$

The third MIF function is simple extension of the PSC function where the spectral correlation is normalized to obtain the spectral coherence. The application of spectral coherence is better when there are large differences in signal energy in various frequency bands. Moreover, the resulting scalar value is normalized to the 0–1 range. The method becomes represented as MID (PSCoh) and is defined as:

$$MID(PSCoh)_{\Delta f}(f;\alpha)=\Delta f\left\langle \frac{SC_{x}^{\alpha}\left(f+\frac{\alpha}{2}\right)}{\sqrt{SC_{x}^{0}\left(f+\frac{\alpha}{2}\right)SC_{x}^{0}(f)}}\frac{SC_{x}^{\alpha}\left(f-\frac{\alpha}{2}\right)}{\sqrt{SC_{x}^{0}\left(f-\frac{\alpha}{2}\right)SC_{x}^{0}(f)}}\right\rangle \tag{5.29}$$

The fourth proposal of the MIF function uses information directly contained in the signal envelope spectrum without the need of the filtering. This idea was triggered by the concept of Protrugram presented in the previous chapter. In this approach kurtosis of the envelope spectrum of the filtered signal can be used as an indicator of modulation intensity. Thus, the method is represented as MID(ESK) and is defined as:

$$MID(ESK)_{\Delta f}(f,\alpha)=\frac{\left\langle \left| E_{\Delta f}(f;\alpha)-E_{\Delta f}(f;\alpha)\right|^{4}\right\rangle_{f}}{\left\langle \left| E_{\Delta f}(f;\alpha)-E_{\Delta f}(f;\alpha)\right|^{2}\right\rangle_{f}^{2}}-3 \tag{5.30}$$

Integrated MID

The Modulation Intensity Density returns the 2-D map containing all the modulations in the input signal. Such a map presents a lot of interesting and useful information, but its interpretation requires much more effort than of a typical 1-D spectrum. The 2-D map is very helpful in searching for transfer paths, but when these are known a simplified tool can be used. For the application in condition monitoring it will be more convenient to create a 1-D plot.

To achieve that the MID map can be simply integrated across the carrier frequency f axis. It will still present the modulating frequencies α, but without distinction between the signal sources or modulating frequencies. The method is called *IMD* which stands for *Integrated MID* and is defined as:

$$IMD_{f_1}^{f_2}(\alpha;\ \Delta f) = \int_{f_1}^{f_2} MID_{\Delta f}(f, \alpha)df \tag{5.31}$$

It can be seen that the IMD is the result of integration of MID in the frequency range $f_1 \ldots f_2$. The examples of the IMD application are given in the next chapter.

5.5.3 Case Study—Simulated Second-Order Cyclostationary Signal

The modulation intensity distribution will be first presented on a simulated second-order modulation signal. The simulated signal was designed to be similar to vibration generated by a rolling element bearing with an outer race local fault. The repetition rate of excited impulses was 80 Hz and their amplitudes were randomly modulated. The carrier signal was stationary Gaussian distributed around 4 kHz. In order to achieve clear results no impulse spacing jitter or additional noise was added. The signal waveform is presented in Fig. 5.48.

Since the signal is second-order cyclostationary, the method MID(PSC) was selected. The PSC abbreviation means that the Modulation Intensity Function (MIF) is the product of Spectral Correlations. The method still uses three separate bandwidths and applies logical "and" operation, but in the following way:

$$\mathrm{MID(PSC)}_{\Delta f}(f;\ \alpha) = \langle x^*_{\Delta f}(t;\ f - \alpha)x_{\Delta f}(t;\ f)e^{-j2\pi\alpha t}\rangle_t \langle x^*_{\Delta f}(t;\ f)x_{\Delta f}(t;\ f + \alpha)e^{-j2\pi\alpha t}\rangle_t \tag{5.32}$$

The resulted modulation intensity will reach non-zero values if both pairs of factors return non-zero correlation which is the signature of the second-order modulation.

The MID map generated for the investigated signal is presented in Fig. 5.49. Fundamental repetition rate of impulses is clearly visible at the cyclic frequency α 80 Hz as the top horizontal line. Next, three harmonics are visible below. The carrier

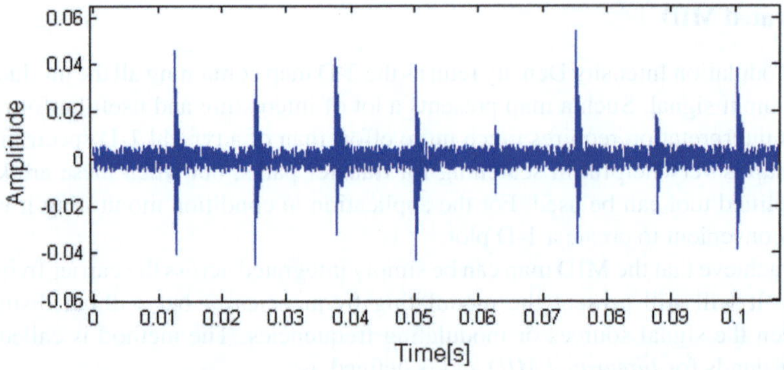

Fig. 5.48 Broadband carrier modulated by periodic impulses with randomly varying amplitudes (reprinted from [16] with permission from Elsevier)

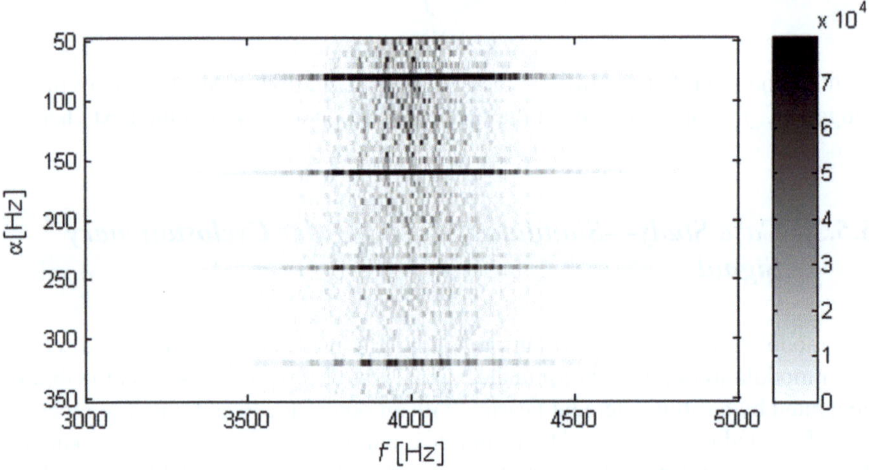

Fig. 5.49 MID map of the simulated signal. Lines parallel to the x axis represent the first four harmonics of the second-order component modulation carried by broad-band noise around 4 kHz (reprinted from [16] with permission from Elsevier)

signal was correctly found to be present in a broadband around 4 kHz. It is important to note that FFT will not detect the modulating impulses as the carrier is a purely random signal. Another interesting feature is the increased noise level in between the detected harmonics in the carrier area. The reason for such a behavior is non-ideal characteristics of the filters used for the MID algorithm and existence of the background noise in the signal.

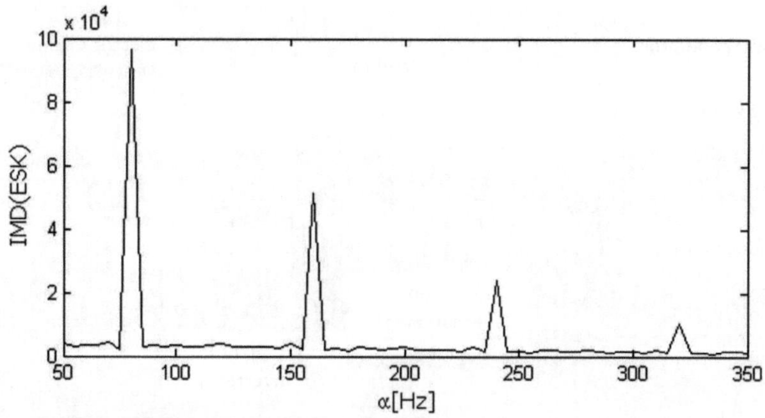

Fig. 5.50 MID from Fig. 5.49 integrated over carrier frequencies between 3–5 kHz (reprinted from [16] with permission from Elsevier)

When we are only interested in revealing the modulating frequencies, the Integrated MID may be used for this purpose. Figure 5.50 presents the IMD for the map presented in Fig. 5.49 and it only shows the distribution along the vertical axis, i.e. the cyclic frequency. The four harmonics of the fundamental 80 Hz frequency are clearly visible.

5.5.4 Case Study—REB Fault Development on the Test Rig

The second case study presents the application of the results of the MID algorithm to the real data from a test rig. The rig contained a faulty REB which had a seeded fault in five severity stages. Figure 5.51 presents the test rig used for the experiment. The vibration sensor was mounted in between the bearings deliberately, as it increased and complicated the transfer path between the faulty bearing and the sensor. It can also be expected that more structural resonances were excited thus making the signal more complex and harder to analyze.

The experiment was executed under stationary conditions in which both rotational speed and load were constant, respectively at 2700 rpm and 100 W. For each fault level 10 s of signal was recorded at 25 kHz. The MID maps were calculated in the frequency range from 1 to 12 kHz. The maximum α frequency was set up to 300 Hz (as the characteristic frequency of the outer race fault was 83 Hz) and the Δf parameter was set to 10 Hz. As a measure of the modulation intensity factor two functions were applied. First, the kurtosis of the envelope spectrum was used, i.e. MID(ESK). Next, the results obtained with MID(PSCoh) were used. Figure 5.52 shows the set of MID maps. The top plot (a) represents the signal from a healthy bearing. The following plots represent signals from the bearing in increasingly severe

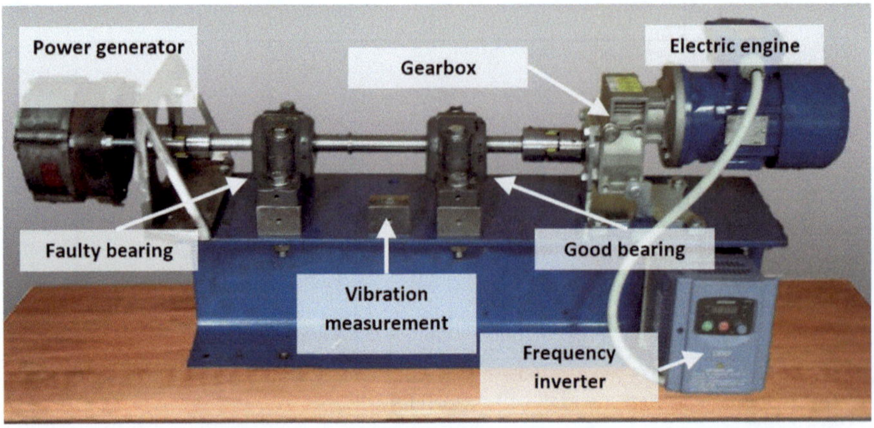

Fig. 5.51 Test rig used in the experiment (reprinted from [16] with permission from Elsevier)

fault conditions. The first signs of failure can be (barely) visible already on the plot (b). The plot (c) shows the first harmonic of the BPFO (83 Hz). The two last plots show increasingly strong horizontal lines representing the fundamental of BPFO and its two harmonics. During the growth of the fault one can observe a change of carrier frequencies. They tend to grow and to some extent—move. After MID, IMD plots were prepared for all the signals and are presented in Fig. 5.53.

For comparison, the Spectral Coherence function was used for the second analysis. Figure 5.54 presents the resulting MID(PSCoh) maps and Fig. 5.55—resulting IMD(PSCoh) plots.

Both results are quite similar, though MID(PSCoh) detected BPFO already at the (b) stage, whereas IMD(ESK) showed non fault related artefacts at frequency around 250 Hz. Differences are better analyzed on IMD figures. The artefacts in MID(ESK) contain disturbances around 4 kHz and 10 kHz and were caused by electrical interferences. The MID(SCoh) seems to be more robust and since the double logical conjunction it is more selective on the carrier frequency axis.

5.6 Instantaneous Circular Pitch Cyclic Power

5.6.1 Overview

Low-speed planetary gearboxes are of special interest as a part of the drivetrain of a geared wind turbine because of large cost of their repairs. Such repairs easily exceed 500 thousand EUR total. However, vibration-based condition monitoring of such gearboxes is challenging and has not reached its full potential yet. The first challenge comes directly from the fact that they operate under non-stationary load

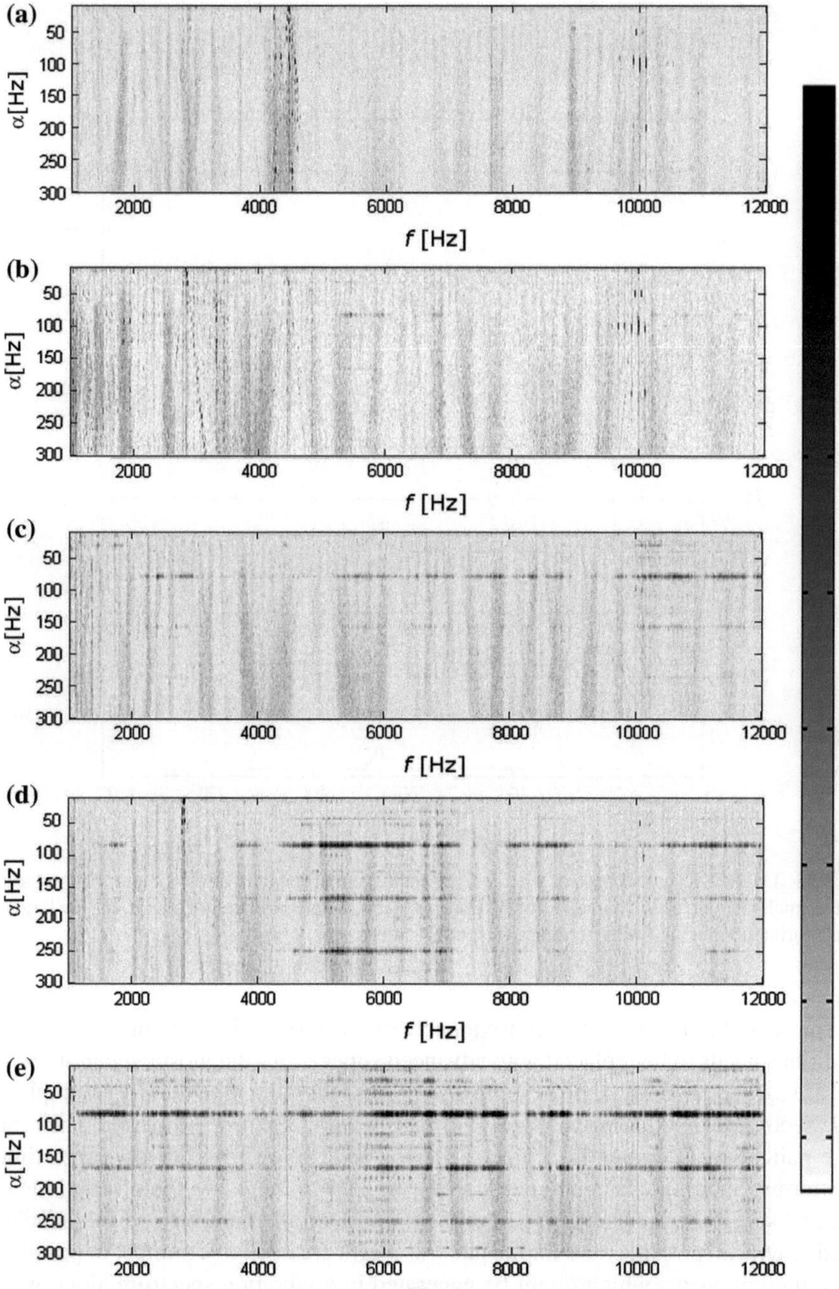

Fig. 5.52 MID(ESK) calculated for signals from each stage of the fault development: **a**—no fault, **b**—the smallest fault, **e**—the largest fault (reprinted from [16] with permission from Elsevier)

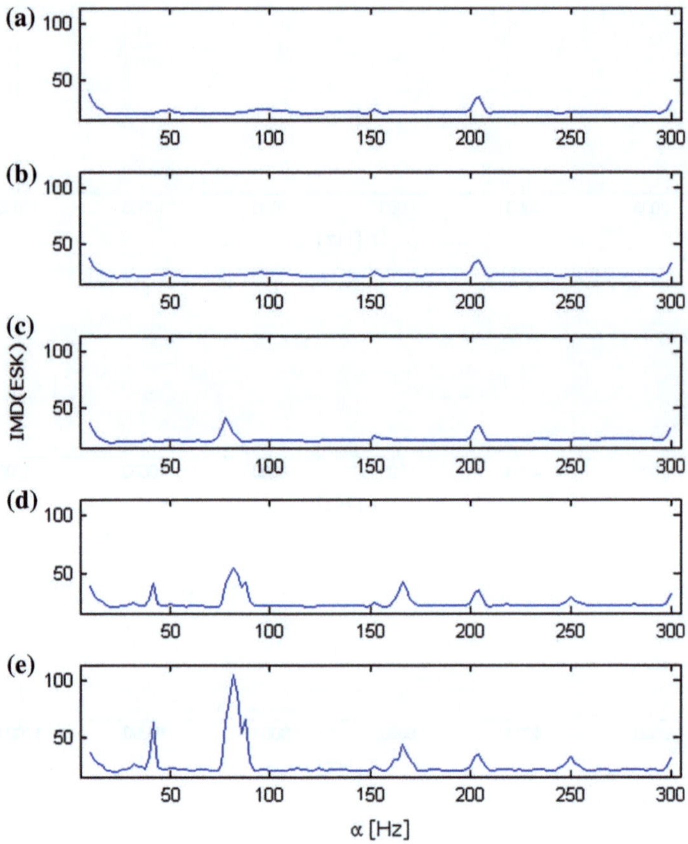

Fig. 5.53 IMD(ESK) calculated for signals from each stage of the fault development: **a**—no fault, **b**—the smallest fault, **c**-first harmonic of BPFO are clearly visible, **d**-developed fault, **e**—the largest fault (reprinted from [16] with permission from Elsevier)

and speed in a broad range. As a consequence, simple monitoring assuming stationary conditions needs to be replaced with advanced supervisory diagnostic techniques as well as sophisticated signal processing algorithms including high fidelity resampling [17]. Secondly, the hunting tooth (HT) design which maximizes the time when the same pair of teeth comes in contact calls for acquisition of relatively large time waveforms. Such signals in practice last at least a few minutes, so some variations in load and speed are inevitable. Thirdly, the number of all characteristic frequencies together with all possible modulation patterns might create a large number of possible spectral components which might be generated in a vibration spectrum along with their harmonics. Finally, it is always worth remembering that generally, any invasive measurement techniques are prohibited on wind turbines; therefore, methods taking advantage of inner sensors are still in a laboratory phase.

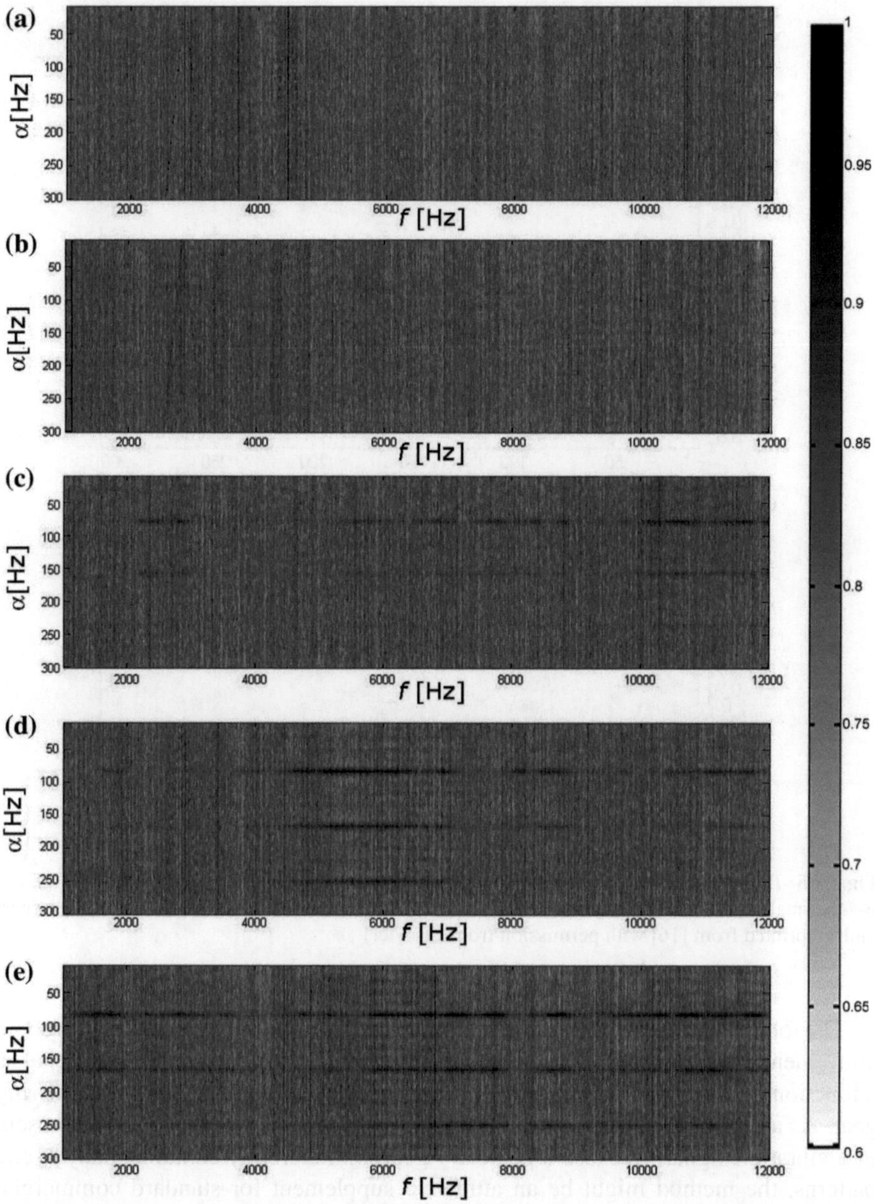

Fig. 5.54 MID(PSCoh) calculated for signals from each stage of fault development: **a**—no fault, **b**—the smallest fault, **c**-first harmonic of BPFO are clearly visible, **d**-developed fault, **e**—the largest fault (reprinted from [16] with permission from Elsevier)

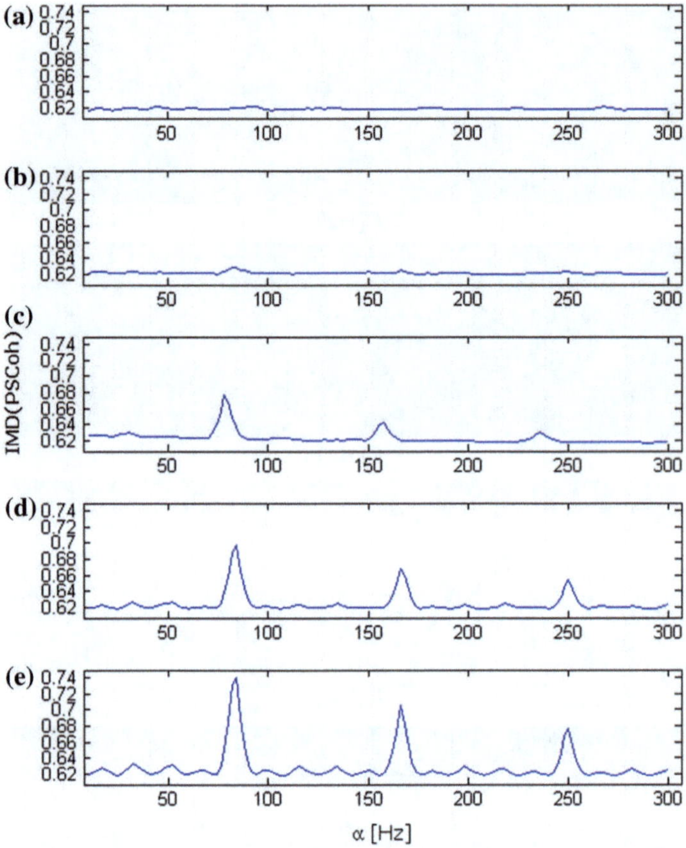

Fig. 5.55 IMD(PSCoh) calculated for signals from each stage of fault development.: **a**—no fault, **b**—the smallest fault, **c**-first harmonic of BPFO are clearly visible, **d**-developed fault, **e**—the largest fault (reprinted from [16] with permission from Elsevier)

The following chapter presents a method which shows some potential to overcome abovementioned obstacles. This method illustrates the power of a vibration signal as a function of the contact angle of subsequent pairs of teeth of the planet and the ring gear. As a result, the method is capable of illustrating some cyclic patterns present in a vibration signal. Because a planetary gearbox inherently contains many cyclic patterns, the method might be an attractive supplement for standard commercial condition monitoring systems. The original paper with description of the method was published by Jablonski and Barszcz [18].

5.6.2 Method Description

ICPCP takes a standard vibration signal along with a reference speed signal. The vibration signal needs to be resampled according to the carrier frequency and the number of teeth of the ring gear in such a way that the resampled signal has got an integer number of samples per carrier rotation, and each resampled signal fragment corresponding to single carrier rotation has to be an integer multiplication of the number of the teeth on the ring gear. If these requirements are fulfilled it is possible to divide the signal into angles corresponding to a single teeth on the ring gear. Next, the resampled signal of the total length of M samples is divided into subsequent fragments related to carrier rotations as illustrated in Fig. 5.56.

A new number of samples of single carrier rotation, K, is calculated by dividing the total number of samples in the resampled signal by the total number of full carrier rotations. If necessary, the signal needs to be truncated. Next, each fragment K is divided with respect to the number of the teeth of the ring gear Z_r, thus creating subsets, each of relatively small length L. From each fragment L, a scalar energy-related value like power or RMS is calculated and is replaced by a single scalar value. Finally, this set of the scalar values of the length equal to $L * K$ is arranged into a matrix form with respect to carrier rotations on one dimension and the ring gear as the other dimension. The concept is illustrated in Fig. 5.57.

As shown in Fig. 5.57, each L sample long fragment of the signal is replaced by the amplitude value representing the energy stored in the time waveform when each tooth is in contact. In this way, an original time waveform is replaced with a 3-dimensional map. The exemplary map is illustrated in Fig. 5.58.

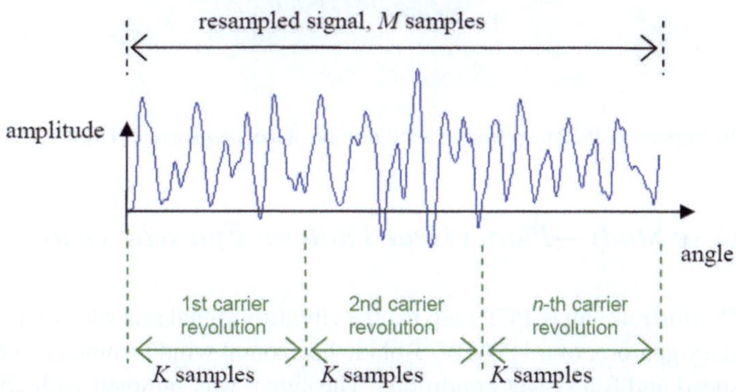

Fig. 5.56 Segmentation of the resampled signal according to carrier rotations [18]

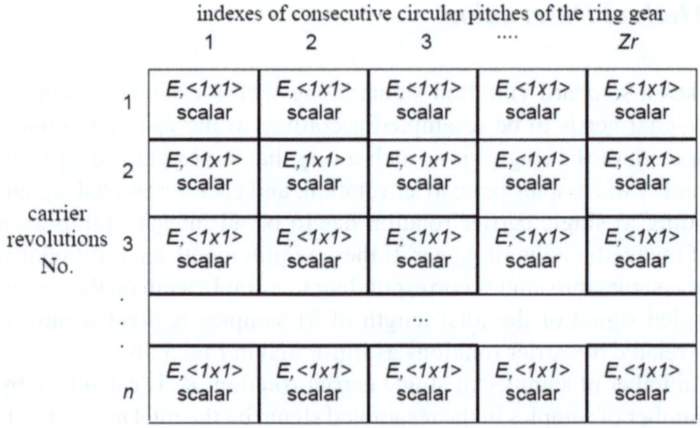

Fig. 5.57 Illustration of a fragmented and processed vibration signal rearranged into a matrix form [18]

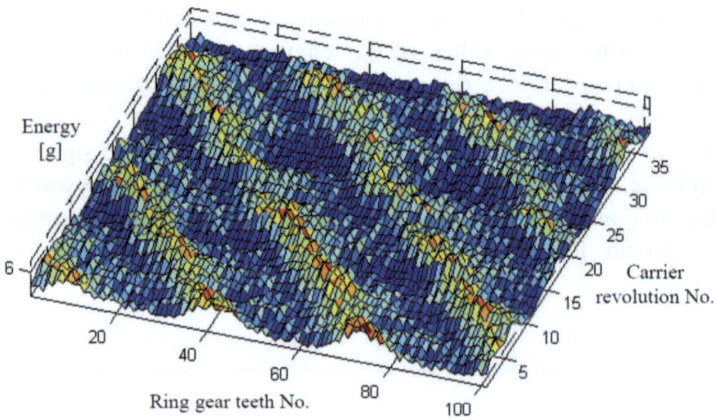

Fig. 5.58 Exemplary ICPCP map from a wind turbine planetary gearbox [18]

5.6.3 *Case Study—Planet Gear Fault on Epicyclic Gear*

Figure 5.58 illustrates an ICPCP map from a vibration signal recorded on the housing of a planetary gearbox of a 1.5 MW, 3-blade horizontal wind turbine recorded in the nominal speed and full power conditions. The signal was sampled with 25 kHz and lasted 300 s. The acquisition unit was a 16 bit continuous acquisition system.

The figure shows clear patterns of amplitude modulations of each planet passing under a vibration sensor which was placed on the housing of the planetary gearbox. Since this is the only pattern visible on the map it might be concluded that no other cyclic patterns are generated or at least detected and recorded by an accelerometer within this gearbox. Moreover, the peak-to-peak version of the ICPCP map might detect and illustrate the pattern of low-energy short-duration time pulse components.

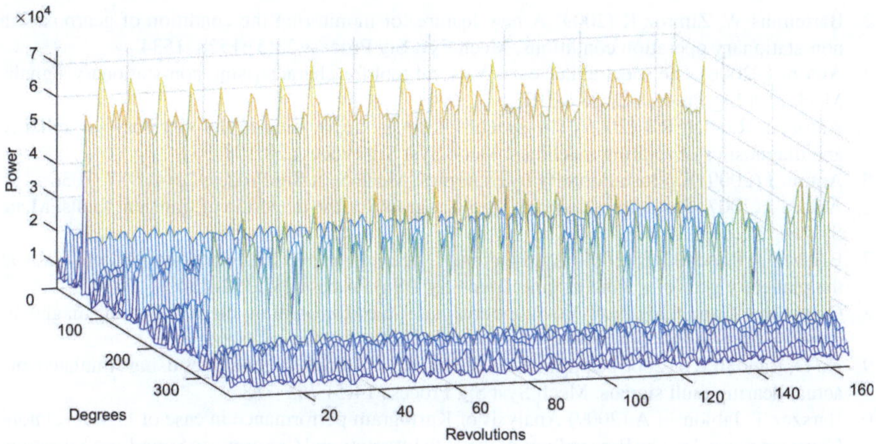

Fig. 5.59 ICPCP map from a test rig signal

5.6.4 Case Study—Planet Tooth Fault on Test Rig Gear

Another case study shows the result of an experimental setup where the capabilities of the ICPCP map were studied on manually induced faults on an academic test rig. Figure 5.59 illustrates an ICPCP map where a single planet tooth was partially removed.

As illustrated in Fig. 5.59, the ICPCP shows two modulation cyclic patters. The first pattern corresponds to planet rotation around the sun gear and is visible as a pair of straight edges. The other pattern corresponding to planet rotation around its own axis is clearly illustrated as cyclic amplitude modulation of the former modulation patters, i.e. straight edges. In this way, the ICPCP enables detection and identification of multi-modulation patterns from a single 3-D figure.

The ICPCP method falls into a category of supervised advanced diagnostic techniques. Other methods using this approach can be found in [19–21]. It is worth mentioning that the major novelty of ICPCP comes from the fact that it does not use any averaging apart from calculating a scalar energy estimator, whereas other methods use some kind of averaging which is in favor of a selected cycle, but simultaneously is detrimental to all the other cycles. As a consequence, for multi-fault scenarios other methods require generation of separate figures.

References

1. Zimroz R, Bartelmus W, Barszcz T, Urbanek J (2014) Diagnostics of bearings in presence of strong operating conditions non-stationarity—A procedure of load-dependent features processing with application to wind turbine bearings. Mech Syst Sig Process 46:16–27

2. Bartelmus W, Zimroz R (2009) A new feature for monitoring the condition of gearboxes in non-stationary operation conditions. Mech Syst Sig Process 23(5):1528–1534
3. Antoni J (2006) The spectral kurtosis: A useful tool for characterising non-stationary signals. Mech Syst Sig Process 20:282–307
4. Antoni J, Randall RB (2006) The spectral kurtosis: application to the vibratory surveillance and diagnostics of rotating machines. Mech Syst Sig Process 20:308–331
5. Antoni J (2009) Cyclostationarity by examples. Mech Syst Sig Process 23(4):987–1036
6. Antoni J (2007) Fast computation of the kurtogram for the detection of transient faults. Mech Syst Sig Process 21:108–124
7. Barszcz T, Randall RB (2009) Application of spectral kurtosis for detection of a tooth crack in the planetary gear of a wind turbine. Mech Syst Sig Process 23:1352–1365
8. Courrech J, Gaudel M (1998) Envelope analysis—the key to rolling-element bearing diagnosis. Bruel & Kjaer Application Notes
9. Ho D, Randall RB (2000) Optimization of bearing diagnostics techniques using simulated and actual bearing fault signals. Mech Syst Sig Process 14(5):763–788
10. Barszcz T, Jablonski A (2009) Analysis of Kurtogram performance in case of high level non-Gaussian noise. In: The Proceedings of the 16th International Congress on Sound and Vibration, Krakow, Poland, 5–9 July 2009
11. Barszcz T, Jablonski A (2011) A novel method for the optimal band selection for vibration signal demodulation and comparison with the Kurtogram. Mech Syst Sig Process 25(1):431–451
12. Gardner W (1994) Cyclostationarity in Communications and Signal Processing. IEEE Press, New York
13. Gardner W (1990) Introduction to random processes. McGraw-Hill, New York
14. Serpedin E, Panduru F, Sari I, Giannakis GB (2005) Bibliography on cyclostationarity. Signal Process 85(12):2233–2303
15. Urbanek J, Barszcz T, Uhl T (2012) Comparison of advanced signal-processing methods for roller bearing faults detection. Metrol Meas Syst 19(4):715–726
16. Urbanek J, Antoni J, Barszcz T (2012) Detection of signal component modulations using modulation intensity distribution. Mech Syst Sig Process 28:399–413
17. Villa LF, Renones A, Peran JR, de Miguel LJ (2011) Angular resampling for vibration analysis in wind turbines under non-linear speed fluctuation. Mech Syst Sig Process 25(6):2157–2168
18. Jablonski A, Barszcz T (2012) Instantaneous circular pitch cyclic power (ICPCP)—a tool for diagnosis of planetary gearboxes. KEM 518:168–173
19. Belsak A, Flasker J (2006) Method for detecting fatigue crack in gears. Theor Appl Fract Mec 46(2):105–113
20. Forrester BD (2001) Method for the separation of epicyclic planet gear vibration signatures, U.S. Patent 6,298,725
21. Maczak J (2009) Local meshing plane as a source of diagnostic information for monitoring the evolution of gear faults. In: Proceedings of the 4th world congress on engineering asset management, Athens, Greece, 28–30 Sept '
22. William A. Gardner, Antonio Napolitano, Luigi Paura, (2006) Cyclostationarity: Half a century of research. Signal Processing 86 (4):639–697
23. Nandi A (ed) (1999) Blind estimation using higher-order statistics, Springer

Chapter 6
Further Research

6.1 Varying Operational Conditions

Historically, the first condition monitoring systems were designed for large machinery, e.g. steam and gas turbines used in power generation. The only non-stationary periods which were analyzed carefully were the transient states. For large rotating machines with journal bearings such an analysis provides the most important information about the shaft line resonances, bearing stability and clearances. Since there was a relatively small number of very costly assets, it was not a problem to install expensive condition monitoring systems and cover skilled labor costs. Still, apart from the aforementioned run-ups and coast-downs, there was not much interest in non-stationary operation analysis.

When wind turbines (WTs) started to grow in installation numbers, there were attempts to install vibration based condition monitoring systems. The first systems were simply condition monitoring systems (CMS) ported from the other machinery types, e.g. heavy duty gearboxes in power generation. Highly variable operating conditions were not taken into account. The application of standard CMS resulted in high false alarm rates. There were frequent cases when wind turbines with a supposed design life of 20 years were failing in five. The data was then collected and the community started to comprehend the specifics of the wind turbines as the object of vibration monitoring. Soon, it became clear that the main difference between wind turbines and other "typical" machinery is the load spectrum. For wind turbines the load follows wind which has inherently random nature and generates much more stress and fatigue in the drivetrain components.

This new field required new research and development efforts. It was a new field of research; however, varying operational conditions had been analyzed before for machines in some industries. Probably the most important direction of research was the helicopter gearboxes, especially the main epicyclic gearboxes and tail gearing. These are complex and high power machines where failure often results in loss of life or severe injuries. Moreover, they operate in a very dynamic environment, where

© Springer Nature Switzerland AG 2019
T. Barszcz, *Vibration-Based Condition Monitoring of Wind Turbines*, Applied Condition Monitoring 14, https://doi.org/10.1007/978-3-030-05971-2_6

drivetrain load is the result of flight envelope and atmospheric conditions. For this type of machinery there was available a way to simplify the condition monitoring problem. The case of helicopters was simplified by inspecting the dynamic behavior in the known conditions. An aircraft is a controlled machine, so it is possible to operate before flight at a known rotational speed or torque. Such a state can be kept for a while, thus creating quasi-stationary conditions. Vibration from such a state can be compared much more easily. The approach of a controlled operation cannot be applied to wind turbines, as the operation follows the wind conditions.

After the Variable Operating Conditions were more and more widely recognized as an important research problem, there was a question of how the wind can be described so it can be better understood and can be used as a reliable input to the WT model. One of initial works was published by Barszcz et al. [1] proving that wind has a fractal nature and can be modelled with the Weierstrass function. In the next step, a similar approach was used by Bielecki et al. [2] to model the chaotic load of WT drivetrains. The papers proposed a few measures to estimate the non-stationarities in the wind and load.

The importance of VOC has been widely recognized in the research community since 2010. That year the conference dedicated to this research subject was established under the name "Condition Monitoring of Machinery in Non-Stationary Operations" (CMMNO). The proceedings of this conference were published and created a valuable resource for the interested reader—see Fakhfakh et al. [3], Dalpiaz et al. [4], Chaari et al. [5] and Timofiejczuk et al. [6]. The area of interest ranges from Rolling Element Bearing diagnostics, through modelling of dynamics and fault in gearing systems, signal processing, condition monitoring techniques to experimental and numerical modeling of machine dynamics and data mining methods. Other mechanical systems, e.g. mining machinery, CNC machinery or aircrafts are also covered by various authors. The other valuable resource is the special issue of the Mechanical Systems and Signal Processing journal devoted to condition monitoring of machines in non-stationary operations, edited by Bartelmus [7]. The presented papers covered several directions of research, important for the VOC aspects. Apart from wind power generation problems from other industries were also presented. In many cases, the analysis of the case studies and methods from other machinery types has been very inspiring to the development of fault detection and identification methods in wind turbines. CMMNO is not the only forum to present research results for the non-stationary operating conditions. Among many others, one can mention the Surveillance conference which is organized every two years and contains sessions devoted to non-stationary regimes of operation.

Discussing directions of the further research can be a very wide subject. Therefore, after sketching the background and the popular organizational frameworks, a few examples of the recent developments in the field will be shown. The development of VOC analysis methods is a very broad field. As far as theoretical foundations are concerned, Bachschmid and Chatterton [8] proposed a differentiation between weak and strong non-stationary conditions. They also compared the behavior of linear and non-linear mechanical systems and showed ineffective sources of simulation of non-linear systems. For the monitoring of strong non-stationary machines modelling becomes very inaccurate. Thus, only accurate signal analysis should be used for the monitoring task.

Numerous new methods are being proposed about how to tackle the non-stationarity problem. A relatively new method is the cointegration. The method uses the residual-based control chart approach. It states that it is possible to find such a linear combination of the nonstationary series that can produce a stationary residual. Next, the stationarity of the residual is monitored for a change. Dao et al. [9] proposed to apply this method for the analysis of the process parameters from a faulty 2 MW wind turbine and show that it is capable of detecting an abnormal technical state.

The important problem in VOC is not only varying rotational speed which can be handled by signal resampling but also by process variables, especially output power which is strongly correlated to the vibration amplitude. Many researchers attempt to develop a method which can "rescale" vibration signals to known operating conditions. From this group, a new modelling technique was proposed by Urbanek et al. [10]. His method takes into account the power value which is then used to scale the vibration signal. For the given examples, signals obtained under different conditions can be compared with each other after application of the proposed normalization technique.

Diagnostics of machinery under VOC is not limited to mechanical drivetrains. Electrical machines are also researched for new methods which can be used in varying conditions. It is not the topic of this book, but—as an example—an interesting method was proposed by Stefani et al. [11]. The authors applied demodulation of current signals which can efficiently analyze transient signals from induction machines and detect faults despite high variability of operating conditions.

On parallel to the scientific research, the International Standardization Committee (ISO) is working further on standards which can be used in the analysis of the WT vibrations. This work is an important and long term endeavor, as it requires to undergo a formal process. Up to the current point (April 2018) it resulted in publication of the new chapter of the norm, namely ISO 10816-21[43]. It was described in the Chap. 2, but it is interesting to mention that this document only covers horizontal axis WTs with a gearbox and power between 10 kW and 3 MW, so it is not applicable to modern, larger designs. The part ISO 10816-22 focused on onshore turbines with generators coupled to a rotor without a gearbox is under preparation. The next parts for offshore turbines (both geared and direct drive) are still in the phase of collecting sufficient amount of data. Having said that, one needs to remember that the ISO standard is constrained to overall averaged vibration values and is not suitable for early fault detection which is of great importance in WTs condition monitoring.

6.2 Fault Modelling

Apart from the development of new signal processing tools it is important to understand the physics of mechanical processes inside the drivetrain. To build an understanding of the mechanical system operation in presence of a fault, two directions of the research are pursued. In the first one, real faults are analyzed and measured vibrations are compared to the real fault. The fault is known via direct inspection or endoscopy. This method is the closest one to the real machine, but a common problem is the scarcity of the available data. The number of possible combinations of several faults grows exponentially. Moreover, the data describing the growth of a fault is not available. It is only possible to have such data for a test rig for a very limited number of faults. So, this direction is a costly one and does not provide sufficient amount of data.

The other approach is to increase the understanding of the process how the fault manifests itself in the vibration signal via modelling. Such an understanding can be built by creating numerical models which take a fault as the input, and generate the output vibration signal. The advantage of numerical models is the ability to generate signals for arbitrary location and size of a fault. It is also possible to combine various fault scenarios and obtain data when more than a single fault is present. This can, in turn, generate a large database of cases which can then be used to train a fault detection method. The fundamental problem in this approach is the accuracy of the model. So, advances in fault modelling are an important step towards the enhanced understanding of faults as well as better fault detection and methods of its identification. There is quite a lot of research in this field and it is not the goal of this chapter to present all the examples and only a few will be listed.

Probably the most common research problem to solve is the relationship between the size of the fault, its impact on the drivetrain dynamics and the resulting vibration signal. Yet another connected problem is understanding how the machine structure determines the transfer path and modifies the vibrations. It can also lead to the optimal selection of the location of vibration sensors. Modelling of gear pairs is not a new area and the first models were already developed 30 years ago, as presented by Ozguven and Houser [12]. One of the further attempts to numerically model gear pairs were proposed by Bartelmus [13]. Probably the first model of the gearbox explicitly suited for VOC was proposed by Bartelmus et al. [14]. The explicit detailed modelling of faulty REBs and gears was proposed by Sawalhi and Randall [15], though no VOC was considered in that work.

The first works generated interest in this field, and were followed with several more detailed investigations. Firstly, simpler models were developed for parallel gears, as presented by Khabou et al. [16] for transient operating conditions. Another example of such an approach for the epicyclic gear is the work of Vicuña and Chaari [17]. The authors developed the numerical model of the single stage planetary gearbox which takes into account the variable load. The model was validated by comparing numerical results with the test rig. An example of the research focused on the gear faults can be the work by Tian et al. [18] where different spur gears were investigated for the relationship between the tooth crack size and the vibration signal.

Other measurements, alternative to mechanical vibrations, are being researched and Instantaneous Angular Speed (IAS) has proved to have a potential for fault detection. An interesting solution by Bourdon et al. [19] proposed a simple method to assess the size of a spall on a REB race. The method uses Instantaneous Angular Speed variations and can be used in VOC. In the later work of Gomez et al. [20] IAS was investigated as a possible source to detect bearing faults in geared systems. The work is at an early stage, but the developed FEM numerical model has shown that bearing faults are detectable by the measurement of IAS. Interestingly, the rotational speed measurements can be taken in different locations of the geared system and not only on the bearing housing.

Similarly to signal processing techniques, the drivetrain modelling is a vast area of research and aforementioned works are merely examples of selected directions. The interested reader can reach for more on the subject on modelling of gear pairs and their faults in the recent review of Liang et al. [21].

6.3 Automated Data Analysis

Proliferation of CMS for wind turbines introduced a new situation for the system developers. First, condition monitoring systems were installed on large critical rotating machinery often in power plants, oil refineries and other heavy industry. There was enough manpower to support the machinery with skilled vibration experts. The diagnostic systems provided the expert users with advanced analysis tools which required training and experience to properly analyze the dynamic state of the machine. After thousands of CMS had been installed on wind turbines the situation changed completely. Instead of a few machines for which expert know-how was available there were many machines operating autonomously from which the data was transferred in an automated way. Human work was too expensive to provide manual analysis of the data from each and every wind turbine. The manufacturers of CMS started to develop various tools for automated analysis of the data stream which day and night flowed into their servers. The situation called for automated data analysis methods. Such a situation may at a first glance seem ideal for the development of the automated tools. It was reinforced by dynamic growth of machine learning discipline [also called artificial intelligence (AI)].

On the other hand, there were—and there are—problems to apply the classical machine learning tools for the machinery diagnostics. A classical approach, so called supervised learning, assumes that there is sufficient amount of data describing all the situations we want to classify. The data set is then divided into the training set and the validation set. After the method is trained on the first set, it is checked for performance on the second set. Such an approach protects against so called overtraining when the results are very good for the training set, but will perform poorly on another one. The problem is now in the availability of the data. Huge amounts of the data incoming to the diagnostic centers is in vast majority generated by machines in a good technical state. The faulty data is scarce and covers only a few faults in one location. On the other hand, the number of possible combinations between types of faults, severity level and location of a fault grows exponentially. There is simply not enough data to provide a sufficient basis for the training of AI tools.

There are some workarounds to this problem, though none of them solve the aforementioned obstacles fully. Firstly, the method can only be trained to the healthy state, or at least to the current state. The AI will then detect any deviation from the known state and only notify a human expert to analyze the data if something new is found. Such an approach is often called novelty detection and is a known part of machine learning world. Secondly, there are AI methods which do not require division into the training and testing set and are referred to as unsupervised learning. Still, the performance compared with the supervised learning is significantly lower and often we can only achieve the novelty detection—at best. Since the unsupervised learning does not have descriptions of the technical state associated with the data, there is a significant part of the work which still must be done by a human expert.

Standard machine learning toolkit contains several groups of tools. The most prominent ones are: classification (e.g. Support Vector Machines (SVM) and decision trees), neural networks (with many types and variants including deep learning network which has gained a lot popularity in recent years), Bayesian networks and expert systems. It is too large a task to provide even brief introduction into the subject in this final chapter. The interested reader should refer to one of many machine learning books, like classic Bishop [22]. For readers interested in examples focused on rotating machinery fault detection, there are review papers available, like the work of Liu et al. [23].

Apart from classical machine learning using the vibration data or features calculated from the vibration signals, there are also other methods and a few selected directions with—in the author's opinion—growth potential will be presented here. Apart from the accelerometers described in the Chap. 3, an important source of diagnostic information are the SCADA systems which are always installed on the wind farms. The Artificial Intelligence methods are developed to analyze the time series. This is an area of vast research and numerous papers. Apart from that, the SCADA systems include the error codes which are also a valuable resource of information. Castellani et al. [24] proposed a method to turn the error codes into indicators on the quality of turbine functionality. This is still an early stage of development, but will most probably be followed in the next years.

There is a large number of potential directions of research in the field of automated data analysis. Methods such as neural networks, decision trees or support vector machines are investigated for an optimal tool. Researchers often try to propose a combination of methods to obtain better results. For example, Fedala et al. [25] proposed a combination of angular-based vibration data with SVM to detect and diagnose gear faults in normal and nonstationary conditions. First, the feature extraction is done from speed independent signals and next, SVM is used for automated classification.

Automated data analysis is also applied in less standard ways. One proposal which can be applied to any CMS is definition of data analysis bins and alarm thresholds. As described in the Chap. 2, the standard approach to the condition monitoring in VOC is using operational states (or bins). Data is collected alongside with the information about the process parameters and later classified into a bin. Despite simplicity of the method, the definition of bins is not a trivial task and can significantly influence the data quality in a bin. Jablonski et al. [26] proposed a method of automated procedure for transformation of the PV data into definition of optimal machine operational states. Not only are the ranges of states obtained, but also their number. The method is illustrated with a real pitch-controlled WT case study. Another, somehow related problem, is the definition of the alarm threshold itself. The classical approach uses mean value of the data and later uses standard deviation value to obtain the threshold level. This approach is effective if the base data follows the normal distribution. As shown by Jablonski et al. [27], the vibration features do not follow the normal distribution and the use of other distributions yields much better results. Despite the importance of the problem of definition of bins and thresholds, it is still not fully researched and deserves more attention.

Last, but not least, there is another group of application of automated data analysis. Instead of analyzing the machinery, it can be applied to analyze a vibration signal without a priori knowledge about the machine kinematic. An interesting example was the effort lead by Martin [28]. Signal processing in this project was described by Gerber et al. [29]. The AStrion project performs steps of automated signal analysis. It starts with resampling to suppress non-stationarities due to small variations of speed and sensor validation, so only correct data is taken into account. The method uses Fourier transform and automatically investigates the signal spectrum to identify peaks and patterns on the spectrum. Further, harmonics and sidebands are detected which may be optionally correlated with the kinematic information. AStrion is a powerful tool and can be applied far beyond wind turbine diagnostics.

6.4 Lifetime Prognostics

Prognostics of the machinery lifetime is the "holy grail" of machinery maintenance. All the analysis methods and building the knowledge about the technical state of drivetrain components is finally used to determine the Remaining Useful Life (RUL) of the machine. The complete answer to this question requires a combination of knowledge about signal processing with not only mechanics, but also material science. The first attempts were possible for industries of mass production and involved statistical approach. It was fundamental to move from the reactive maintenance to the preventive maintenance strategy. Sometimes it can be purely based on statistics, but is effective only when there is sufficient amount of data available. The general statistical approach can be found in the work of Phelps et al. [30]. The advantage of this method is that it does not require condition monitoring to be installed; the disadvantage it is that it only works for large populations of identical members and the results can not be held valid if the operating conditions change.

For real life mechanical systems the condition of having sufficient amount of data, including fault cases is often impossible to fulfill. In such cases other approaches can be applied. On the one hand, physical fault and degradation models are being developed. On the other hand—mathematical regression models are developed. The typical approach to the prognostics task is a process which includes a few steps from data acquisition, through feature extraction, severity assessment to RUL prediction. There is a great variety of different approaches and the complete description is beyond the scope of this chapter. A thorough review of prognostic methods can be found in already classical work of Jardine et al. [31]. Another, more recent review was prepared by Lei et al. [32]. The book on the subject was written by Kim et al. [33]. Certainly, non-stationary systems introduce significant difficulties to the prediction task and the interesting review of possible methods was given by Kan et al. [34]. Another review, focused entirely on wind turbine prognostics was given by Leite et al. [35]. As this chapter does not aim to present the complete landscape of the prognostics, only a few aspects will be presented in more detail below.

The example of an approach to the prediction problem is illustrated in the work of Dong et al. [36]. Several methods are combined to create an algorithm for the RUL task. After vibration features are selected for the CM process, there is a need of dimensionality reduction which is performed with the Support Vector Machine (SVM). The variability of the resulting features is still very high, and the Markov model is used to improve the accuracy of the RUL prediction. The method was validated on the stationary data from the test rig.

Any prediction task does not result in exact answers, thus it is important not only to obtain the RUL prediction, but also the uncertainty of the result. As shown by Sankararaman [37], there are different ways to understand uncertainty. Moreover, the prediction task should be analyzed in terms of uncertainty propagation, as the inevitable inaccuracies in the input data will result in inaccuracy of the RUL prediction.

Among the more specific attempts to wind turbine prognostics is the work of Saidi et al. [38]. The subject is the bearing of the WT generator shaft. The authors apply the spectral kurtosis tools to extract signal features and later to trend the degradation. The work is more focused on the feature extraction part. In yet another paper the authors present the integrated approach to generator bearing prognostics—see Saidi et al. [39]. Here the physical model of the bearing spall was integrated with the feature tracking. The combination of physical modelling was also proposed by Leturiondo et al. [40]. The work explicitly assumed the non stationary loading of rolling element bearing. The loads and stresses were calculated with the FEM method. The result of the calculation is the fatigue which is then used for the RUL estimation.

The other research direction proposes analyzing not only the frequency content of the vibration signals, but also its general complexity. Boškoski et al. [41] proposed to analyze the complexity of the envelope combined with Gaussian Process model. The method was validated on the test rig data.

It was mentioned before, that vibration is not the only element used for the estimation of a machine technical state. The parameter which has a large potential for RUL estimation is the amount of particles in the oil. Particle filters are installed as a new part of the CM strategy. An interesting review of this direction of research was presented by Jouin et al. [42].

References

1. Barszcz T, Bielecka M, Bielecki A, Wójcik M (2012) Wind speed modelling using Weierstrass function fitted by a genetic algorithm. J Wind Eng Ind Aerod 109:68–78
2. Bielecki A, Barszcz T, Wójcik M (2015) Modelling of a chaotic load of wind turbines drivetrain. Mech Syst Sig Process 54–55:491–505
3. Fakhfakh T, Bartelmus W, Chaari F, Zimroz R, Haddar M (eds) (2012) Condition monitoring of machinery in non-stationary operations. In Proceedings of the second international conference "condition monitoring of machinery in non-stationnary operations" CMMNO'2012
4. Dalpiaz G, Rubini R, D'Elia G, Cocconcelli M, Chaari F, Zimroz R, Bartelmus W, Haddar M (eds) (2014) Advances in condition monitoring of machinery in non-stationary operations. In: Proceedings of the third international conference on condition monitoring of machinery in non-stationary operations CMMNO 2013
5. Chaari F, Zimroz R, Bartelmus W, Haddar M (eds) (2016) Advances in condition monitoring of machinery in non-stationary operations. In: Proceedings of the fourth international conference on condition monitoring of machinery in non-stationary operations, CMMNO'2014, Lyon, France December 15–17
6. Timofiejczuk PA, Chaari F, Zimroz R, Bartelmus W, Haddar M (eds) (2018) Advances in condition monitoring of machinery in non-stationary operations. In: Proceedings of the 5th international conference on condition monitoring of machinery in non-stationary operations, CMMNO'2016, 12–16 September 2016, Gliwice, Poland
7. Bartelmus W (ed) (2013) Special issue: condition monitoring of machines in non-stationary operations. Mech Syst Sig Process 38(1):1–264
8. Bachschmid N, Chatterton S (2014) Dynamical behavior of rotating machinery in non-stationary conditions: simulation and experimental results. In: Dalpiaz G et al (eds) Advances in condition monitoring of machinery in non-stationary operations. Lecture notes in mechanical engineering. Springer, Berlin, Heidelberg
9. Dao PB, Staszewski WJ, Uhl T (2018) Operational condition monitoring of wind turbines using cointegration method. In: Timofiejczuk A, Chaari F, Zimroz R, Bartelmus W, Haddar M (eds) Advances in condition monitoring of machinery in non-stationary operations. Applied condition monitoring, vol 9. Springer, Cham
10. Urbanek J, Strączkiewicz M, Barszcz T (2014) Joint power-speed representation of vibration features. application to wind turbine planetary gearbox. In: Dalpiaz G et al (eds) Advances in condition monitoring of machinery in non-stationary operations. Lecture notes in mechanical engineering. Springer, Berlin, Heidelberg
11. Stefani A, Bellini A, Filippetti F (2009) Diagnosis of induction machines' rotor faults in time-varying conditions. IEEE Trans Ind Electron 56:4548–4556
12. Ozguven HN, Houser DR (1988) Mathematical-models used in gear dynamics—a review. J Sound Vib 121(3):383–411
13. Bartelmus W (2001) Mathematical modelling and computer simulations as an aid to gearbox diagnostics. Mech Syst Signal Process 15(5):855–871
14. Bartelmus W, Chaari F, Zimroz R, Haddar M (2010) Modelling of gearbox dynamics under time-varying non stationary load for distributed fault detection and diagnosis. Eur J Mech A/Sol 29:637–646
15. Sawalhi N, Randall RB (2008) Simulating gear and bearing interactions in the presence of faults Part. 1. The combined gear bearing dynamic models and the simulation of localized bearing faults. Mech Syst Sig Process 22:1924–1951
16. Khabou MT, Bouchaala N, Chaari F, Fakhfakh T, Haddar M (2011) Study of a spur gear dynamic behavior in transient regime. Mech Syst Sig Process 25(8):3089–3101
17. Vicuña CM, Chaari F (2016) Analysis of a planetary gearbox under non-stationary operating conditions: numerical and experimental results. In: Chaari F, Zimroz R, Bartelmus W, Haddar M (eds) Advances in condition monitoring of machinery in non-stationary operations. CMMNO 2014. Applied condition monitoring, vol 4. Springer, Cham

18. Tian Z, Zuo MJ, Wu S (2012) Crack propagation assessment for spur gears using model-based analysis and simulation. J Intell Manuf 23(2):239–253
19. Bourdon A, Rémond D, Chesné S, André H (2014) Reconstruction of the instantaneous angular speed variations caused by a spall defect on a rolling bearing outer ring correlated with the length of the defect. In: Dalpiaz G et al (eds) Advances in condition monitoring of machinery in non-stationary operations. Lecture notes in mechanical engineering. Springer, Berlin, Heidelberg
20. Gomez JL, Khelf I, Bourdon A, André H, Rémond D (2018) Simplified dynamic model of a wind turbine shaft line operating in non-stationary conditions applied to the analysis of IAS as a machinery surveillance tool. In: Timofiejczuk A, Chaari F, Zimroz R, Bartelmus W, Haddar M (eds) Advances in Condition monitoring of machinery in non-stationary operations. Applied Condition Monitoring, vol 9. Springer, Cham
21. Liang X, Zuo MJ, Feng Z (2018) Dynamic modeling of gearbox faults: a review. Mech Syst Sig Process 98:852–876
22. Bishop CM (2011) Pattern recognition and machine learning. Springer
23. Liu R, Yang B, Zio E, Chen X (2018) Artificial intelligence for fault diagnosis of rotating machinery: a review. Mech Syst Sig Process 108:33–47
24. Castellani F, Astolfi D, Terzi L (2016) Analyzing state dynamics of wind turbines through SCADA data mining. In: Chaari F, Zimroz R, Bartelmus W, Haddar M (eds) Advances in condition monitoring of machinery in non-stationary operations. CMMNO 2014. Applied condition monitoring, vol 4. Springer, Cham
25. Fedala S, Rémond D, Zegadi R, Felkaoui A (2016) Gear fault diagnosis based on angular measurements and Support Vector Machines in normal and nonstationary conditions. In: Chaari F, Zimroz R, Bartelmus W, Haddar M (eds) Advances in condition monitoring of machinery in non-stationary operations. CMMNO 2014. Applied condition monitoring, vol 4. Springer, Cham
26. Jablonski A, Barszcz T, Wiciak P (2014) Non-clustering method for automatic selection of machine operational states. In: Dalpiaz G et al (eds) Advances in condition monitoring of machinery in non-stationary operations. Lecture notes in mechanical engineering. Springer, Berlin, Heidelberg
27. Jablonski A, Barszcz T, Bielecka M, Breuhaus P (2013) Modeling of probability distribution functions for automatic threshold calculation in condition monitoring systems. Measurement 46(1):727–738
28. Martin N (2015) KAStrion project: a new concept for the condition monitoring of wind turbines. In: Proceedings of twelve international conference on condition monitoring and machinery failure prevention technologies CM2015, 2015. Oxford, United Kingdom
29. Gerber T, Martin N, Mailhes C (2015) Time-frequency tracking of spectral structures estimated by a data-driven method. IEEE Trans Ind Electron 62(10):6616–6626
30. Phelps E, Willett P, Kirubarajan T (2001) A statistical approach to prognostics. In: Component and systems diagnostics, prognosis and health management, vol 4389. Bellingham, pp 23–34
31. Jardine AK, Lin D, Banjevic D (2006) A review on machinery diagnostics and prognostics implementing condition based maintenance. Mech Syst Sig Process 20(7):1483–1510
32. Lei Y, Li N, Guo L, Li N, Yan T, Lin J (2018) Machinery health prognostics: a systematic review from data acquisition to RUL prediction. Mech Syst Sig Process 104:799–834
33. Kim NH, An D, Choi JH (2017) Prognostics and health management of engineering systems. Springer
34. Kan MS, Tan A, Mathew J (2015) A review on prognostic techniques for non-stationary and non-linear rotating systems. Mech Syst Sig Process 62–63:1–20
35. Leite GNP, Araújo AM, Rosas PAC (2018) Prognostic techniques applied to maintenance of wind turbines: a concise and specific review. Renew Sust Energ Rev 81(2):1917–1925
36. Dong S, Yin S, Tang B, Chen L, Luo T (2014) Bearing degradation process prediction based on the Support Vector Machine and Markov model, Shock Vib 2014 (Article ID 717465)
37. Sankararaman S (2015) Significance, interpretation, and quantification of uncertainty in prognostics and remaining useful life prediction. Mech Syst Sig Process 52–53:228–247

38. Saidi L, Ben AJ, Bechhoefer E, Benbouzid M (2017) Wind turbine high-speed shaft bearings health prognosis through a spectral Kurtosis-derived indices and SVR. Appl Acoust 120:1–8
39. Saidi L, Ben AJ, Benbouzid M, Bechhoefer E (2018) An integrated wind turbine failures prognostic approach implementing Kalman smoother with confidence bounds. Appl Acoust 138:199–208
40. Leturiondo U, Salgado O, Galar D, Mishra M (2016) Methodology for the estimation of the fatigue life of rolling element bearings in non-stationary conditions. In: Chaari F, Zimroz R, Bartelmus W, Haddar M (eds) Advances in condition monitoring of machinery in non-stationary operations. CMMNO 2014. Applied condition monitoring, vol 4. Springer, Cham
41. Boškoski P, Gašperin M, Petelin D (2014) Signal complexity and gaussian process models approach for bearing remaining useful life estimation. In: Dalpiaz G et al (eds) Advances in condition monitoring of machinery in non-stationary operations. Lecture notes in mechanical engineering. Springer, Berlin, Heidelberg
42. Jouin M, Gouriveau R, Hissel D, Marion-Péra MC, Zerhouni N (2016) Particle filter-based prognostics: review, discussion and perspectives. Mech Syst Sig Process 72–73:2–31
43. ISO 10816-21 (2015) Mechanical vibration—evaluation of machine vibration by measurements on nonrotating parts—Part 21: onshore wind turbines with gearbox (2015) International Standard Organization

MIX
Papier aus verantwortungsvollen Quellen
Paper from responsible sources
FSC® C105338

If you have any concerns about our products,
you can contact us on
ProductSafety@springernature.com

In case Publisher is established outside the EU,
the EU authorized representative is:
Springer Nature Customer Service Center GmbH
Europaplatz 3, 69115 Heidelberg, Germany

Printed by Libri Plureos GmbH
in Hamburg, Germany